Water for Rural Communities

Helping People Help Themselves

John Briscoe and David de Ferranti

The World Bank
Washington, D.C.

Manufactured in the United States of America

First Printing March 1988

The World Bank does not accept responsibility for the views expressed herein, which are those of the authors and should not be attributed to the World Bank, to its affiliated organizations, or to its Board of Executive Directors and the governments they represent.

When this report was written, John Briscoe and David de Ferranti were, respectively, a consultant to, and Chief of, the Policy and Research Division in the Water Supply and Urban Development Department. They are now, respectively, Senior Economist in the Brazil Country Department and Chief of the Population and Human Resources Operation Division in the Southern Africa Department.

Library of Congress Cataloging-in-Publication Data

Briscoe, John, 1948-
Water for rural communities.

Bibliography: p.
1. Water-supply, Rural--Developing Countries. 2. Drinking water--International cooperation. 3. International Drinking Water Supply and Sanitation Decade, 1981-1990. I. De Ferranti, David M. II. Title.
TD327.B75 1988 363.6'1'09172 87-31750
ISBN 0-8213-0964-1

Contents

Acknowledgments

This paper resulted from work performed under the general direction of Anthony Churchill, who at the time was Director of the Water Supply and Urban Development Department, and Ping-Cheung Loh, who succeeded him. In addition to the authors, contributions were made by Anthony Churchill, Joseph Freedman, James Listorti, Edward Quicke, Robert Roche, Arumukham Saravanapavan, Alan Walters, Albert Wright, Fred Wright, and Anthony Yezer.

Constance Paige and Carolyn Tager helped write earlier drafts. Brian Appleton helped edit the final text. Deborah Joichin, Jocelyne Milord and Michael Alloy contributed extensive time and effort to the production of the manuscript. The staff of the World Bank/UNDP Handpumps Project, which is directed by Saul Arlosoroff, provided invaluable advice, information and assistance throughout.

Earlier drafts of the paper were distributed to many agencies involved in the water sector, including UNICEF, the United Nations Development Programme, the World Health Organization, the Danish International Development Agency, the International Development Resarch Centre of Canada, the Kreditanstalt fur Wiederaufbau of the Federal Republic of Germany, and the U.S. Agency for International Development. Useful comments were received from many of these organizations and from colleagues in the regional Water Supply Divisions of the World Bank.

1. What Has Been Done

More than 1,500 million people -- 30 percent of the world's population -- still do not have access to safe water. The consequences for their health, productivity and quality of life, as well as secondary implications for economic development, the role of women and poverty alleviation, are believed to be enormous. Since the 1960s, extensive efforts have been mounted to correct this problem, with increased emphasis since the inauguration in 1981 of the United Nations' International Drinking Water Supply and Sanitation Decade.

Significant advances have been made, and the estimated number of rural people without access has been reduced from about 80 percent of the rural population of developing countries in the 1960s to less than 60 percent at present. But the pace of progress has fallen far short of that hoped for by most governments, donors, and unserved communities. The Decade's goal of universal coverage remains a distant vision.

Other causes for concern have come to light, too. Many improvement projects have proved to be neither sustainable nor replicable. Rapid population growth has made a large task larger. Widespread failures of new water supply systems, mainly from inadequate maintenance, have undercut the gains. A harsh worldwide economic climate, especially hard on the developing countries, which suffer flagging economic growth and high debt burdens, has imposed tighter constraints on the availability of resources for investments in water supply and sanitation.

Large sums are being spent on improvement efforts. Yet, to reach the goal of universal coverage will require much larger sums -- far exceeding what governments and donors have invested so far or are likely to be able to invest in the years ahead. While no one has precise figures, it has been estimated that governments and donors are currently spending about US$1,500 million a year on rural water supply. Estimates of the amounts required to reach universal coverage by 2000 range from three to five times that figure. It is evident that governments and donors can provide only a modest fraction of the resources needed.

During the past several years, the World Bank has examined the sources of past disappointments and has sought to develop a new view of what the guiding principles should be. This work has culminated in three recent major reports. The first, *Community Water Supply: The Handpump Option* (Arlosoroff and others 1987), focuses on the technical and policy issues relating to one of the principal technologies used to supply water to rural people. The second report, *Rural Water Supply and Sanitation: Time for a Change* (Churchill 1987), explores in detail the importance of time savings as a benefit from investments in the rural water supply sector, and the implications for planning and investment in the sector. This report, the third in the series, describes the principles that should guide development of the sector. The major themes of this report are:

- It is vital that all parties involved in efforts to improve community water supply -- government agencies, donors, advisors, community leaders, and residents -- recognize and adhere to the principle that *it is the local people themselves, not those trying to help them, who have the most important role. The community itself must be the primary decision maker, the primary investor, the primary maintainer, the primary organizer, and the primary overseer.*
- The remaining themes stem from this fundamental precept.
- It is crucial to determine what the community really wants and is able and willing to sustain. Local preferences and local ability and willingness to commit resources must be the guiding force in every aspect of improvement efforts, from design through implementation. In particular this includes the choice of technology, whether the choice is for low-cost open wells at one extreme or more costly systems involving individual house connections at the other. If communities choose and can support a higher or lower level of service than that recommended by outside experts, then local will should prevail.
- An integral and essential part of an effective strategy is to mobilize the community's own resources, both financial and nonfinancial. This is necessary to assure that the community truly is in control, that systems remain operating, and that the limited funds available

to the government be directed to wherever they are needed most. A goal of every improvement effort should be to bring closer the day when the community can cover all of the costs of its water service from its own resources. Many communities could and should contribute more now to meeting their costs than they have been expected to in the past.

- The primary role of government agencies and donors must change from that of direct providers and financers of services to that of facilitators.
- As facilitators, government agencies and donors have a variety of vital supportive functions to perform in the areas of training, information dissemination, and technical and managerial assistance. They should also work to reduce barriers to the development of private sector suppliers of equipment, drilling and maintenance services.
- Institutional responsibilities need to be redefined, and initiatives to strengthen institutions need to be reoriented in many countries, so that they are more consistent with the appropriate roles of government agencies, donors, and communities.

For some, these ideas -- especially the implication that the disadvantaged rural people should pay for improved water services -- will seem a hard-hearted, commercial approach. The evidence, however, is now in: the worst possible approach is to see poor people as having no resources. Treating the people as the objects (not the subjects) of development inevitably leads to unsustainable subsidies, which reach primarily those of greater influence and least need, and to malfunctioning and restricted services which leave the lot of the poor unchanged. Promises of free service for all too often result, in practice, in some service for a few and little or none for most.

There can be no escaping the fact that in the severely resource-constrained conditions that most developing countries will continue to face in the decades ahead, the best resource available to people in underdeveloped communities is themselves. Other resources -- from government agencies and aid donors -- are too likely, in practice, to be inadequate, inappropriate, or unreliable from the community's perspective, as well as unsustainable from the national perspective. Such resources can help, but they cannot take the place of the resources of the community itself. Everyone involved at every level in efforts to improve water supplies must face up to this. There must be an awareness, in short, that local people are their own best asset and their own most promising mechanism for bringing about change.

The potential advantages of proceeding from this outlook, instead of from the older top-down approaches are considerable. Improvement efforts are more likely to meet felt needs. Investments are more likely to be well spent. New projects are more likely to be completed and kept in service, since they will be locally led and will depend less on outside funds, materials, and personnel that may not be forthcoming when needed. As a result, the country's overall resources are likely to be more efficiently utilized.

No less important, more communities are more likely to get safe water sooner and, because of that, water sector policies and programs will be better instruments to help alleviate poverty, improve the lot of women, and increase equity. As communities are encouraged and supported to do as much for themselves as they can from their own resources, the limited funds available from government sources and donor aid can be conserved and used wherever they are needed most, whether to assist the poorest communities or to facilitate faster economic growth.

There is nothing wholly new in these ideas and nothing particularly complicated. In a way, they represent a return to concepts that the residents of developing communities might say have been obvious to them all along.

To appreciate what led to the recent thinking at the World Bank regarding this sector, it is important to be aware of certain facts about: the benefits of community water supply improvements; the present situation and prospects for future progress; advances in low-cost technology, and the World Bank's experience with the sector.

The Benefits of Improving Community Water Supply

Improving water supplies has been a high priority activity for most developing country governments, donor agencies, and communities for many years now. This emphasis has occurred after considerable reflection about whether two fundamental conditions are satisfied, conditions that are crucial in making appropriate choices in any sector. The first is whether the benefits of water supply improvements exceed the costs. The second is whether some form of public action -- by the community or higher level authorities -- is required or whether private individual initiatives are sufficient.

What has been learned is that the answers to these questions depend on the particulars: on the particular improvement plan proposed; on the particular types and severity of market failures, distortions, and government interventions that may exist in the water supply sector and the markets for other goods and services; on the particular objectives of communities and higher level authorities; and on the particular stance they take on whether to alleviate poverty immediately and redistribute wealth or, alternatively, to emphasize the acceleration of economic growth. Much has been learned, too, about the costs of the various technological options and about the importance of counting all the costs.

Most important, the benefits are now understood better. Water supply improvements can have several different kinds of benefits: those related to health, time savings, other direct effects (better reliability of service, more convenient hours of operation, and better water quality in terms of taste, clarity, and odor), and indirect effects (for example, on poor people's incomes, if, as a result of improved health and time savings, they become more productive).

The effect on *health* has been hypothesized by some to be of vital significance for overall development. Others have questioned this view, pointing out that many water projects have not produced convincing evidence of health gains. Recent reexaminations of the data have helped to clarify the issue (Box 1). There can now be no doubt that water supply improvements do have considerable potential to help improve health. But if that potential is to be realized, many other changes must be brought about in rural communities besides simply installing new hardware. The hardware components of water projects are only one link in a long chain. The other links involve changing hygiene habits and other factors and can require actions ranging from providing better education to promoting public health programs. If any one link is missing, health indicators may not improve. But this does not mean that investing in individual links is futile. Where it is not possible to upgrade all links simultaneously, one must take a longer view, proceed step by step, and not expect to see large health improvements until the last step has been completed.

The *time savings* benefit results from delivering water closer to where people live and work. For vast numbers of communities, the task of getting water, including walking to and form the source and waiting in queues, is still very time consuming. In some places, families spend several hours a day getting water. Improvement projects have reduced the time substantially. In a village on the Mueda Plateau in Mozambique, an improvement project reduced the average time that women spend collecting water to 25 minutes; in a similar village that still relied on traditional water sources, the average time was more than two hours. Saving time also means saving energy, since carrying water-laden containers long distances is strenuous work. This is an important consideration for people living on limited calorie-protein intake and dependent on their own physical labor in fields or cottage industries.

In addition, the time saved can be used for directly productive activities or a variety of other household tasks. A gain of some kind in people's well-being is achieved regardless of whether the time is used to cultivate crops, tend a home garden, trade in the market, keep small livestock, care for children, get more education, or even simply rest and gain strength for the next day.

The role of women in rural communities and the contributions that improved water supply can make in the lives of women are important. The time saved is the time, typically, of the women in the household, for it is usually they who fetch the water. Their time and energy are among the household's most important resources and are vital for the health and well-being of the household's members, for the rearing of its children, and for the success of its productive activities.

The other *direct benefits* of water supply improvements -- reliable service, convenient hours of operation, taste, clarity, and odor -- are often treated as unimportant in planning efforts, in part because they are difficult to quantify. In fact, though, in some situations, they have been more crucial than the health or time-savings benefits. In some communities, for example, the residents have been willing to invest resources in putting a new well right next to a traditional unimproved supply even though there may be no health gains or time savings.

The *indirect benefits* which are related to promoting economic development and poverty alleviation, are easier to speculate about than to document. Insofar as water supply improvements help communities to become healthier and result

Box 1. Health Effects of Improved Water Supply and Sanitation

The declaration of the International Drinking Water Supply and Sanitation Decade and the formulation of current strategies for rural water supply and sanitation are predicated on the assumption that use of improved facilities will improve health substantially. However, for a number of reasons, efforts do demonstrate and measure the link between improvements in water and sanitation and benefits to health have not always succeeded.

One problem is the variety of ways in which better water supply and excreta disposal facilities actually affect health. Some diseases (such as typhoid and cholera) are often *water-borne* and are thus affected mainly by the bacteriological quality of the drinking water; other diseases (such as shigellosis and trachoma) are usually *water-washed* and so depend on the quantity of water used for personal hygiene; another group (including schistosomiasis) is *water-washed* and requires that direct contact between people and infected water should be minimized; and some diseases (such as sleeping sickness and river blindness) are transmitted by *water related insect vectors* which breed in or near water sources, and require that populations be located away from the water sources. Another difficulty in evaluating health improvements is separating the effects of water supply and sanitation interventions from other factors that may contribute to the spread of disease.

Despite these and other serious methodological problems, it is possible to compile an overall picture of the effect of improved water supply and sanitation on one of the more important health problems -- diarrheal disease in young children. Box Table 1.1 shows the results of a recent literature review by the World Health Organization's Diarrheal Diseases Control Program. Although some studies demonstrated no improvement in health, the review showed that typically diarrhea morbidity is reduced when water supply and sanitation are improved and that reductions in mortality from diarrheal diseases are somewhat greater. The effects are by no means negligible, but they are substantially lower than predictions that are commonly used to justify investments in water and sanitation projects.

On both theoretical and empirical grounds, the conclusion is that improved water supply and sanitation are necessary, but not always sufficient, conditions for improved health. More specifically, the effect of improved water supplies is likely to be greatest when there are concomitant improvements in excreta disposal practices and other personal hygiene behaviors. The corollary is that the changes necessary for an effect on health are most likely to occur where there is a real demand for improved excreta disposal facilities, such as in densely-populated areas. Furthermore, it is in the urban fringes and more densely populated rural areas where the effect on health of improved sanitation is likely to be greatest.

Ten years ago, a World Bank expert panel concluded that "long-term longitudinal studies of large size and expense are probably the only means through which there is any chance of isolating a specific quantitative relationship between water supply and health." Recent developments suggest that case-control and cross-sectional methods have promise as ways of assessing such a relationship rapidly, validly, and at reasonable cost. However, such evaluations would be appropriate only where, after an assessment of direct economic benefits and willingness to pay, it is uncertain whether a project should proceed and where it is considered likely that the improvements will have a substantial effect on health.

Box Table 1.1. Typical Effects of Improved Water Supply and Sanitation Conditions on Diarrheal Morbidity

Conditions	*Median reduction in diarrhea morbidity (percent)*
Improved water quality	16
Improved water availability	25
Improved water quality and availability	37
Improved excreta disposal	22

Source: Esrey and others (1985).

in time savings, they make it possible for people to be more productive, and increased productivity gives rise to higher levels of income.

Poverty alleviation is aided in that better health, saved time, greater productivity (and hence income), and -- last but not least -- the better water itself are among the requirements for raising the living standard of the poor. Water projects, like education and health-care projects, are among the most common kinds of direct initiatives to alleviate poverty that countries can promote.

The Present Situation and Future Outlook

The number of people without access to safe water is still very large -- an estimated 1,500 million, as quoted earlier. From figures reported by government agencies to the World Health Organization, it has been estimated that in 1983 nearly 60 percent of rural people in developing countries did not have access to safe water (Table 1).

The actual number unserved was probably substantially greater than 60 percent, since the reported coverage figures often overstate those who truly have reasonable access and choose to use the improved service. Recent in-depth assessments in East and West Africa revealed that the population actually using improved water supply schemes was only one-third of the population reported to have access. A survey in India showed that only about one-half of the villagers nominally served by handpumps actually used them.

Table 1. Rural Water Supply Coverage, 1983

Region	*Percentage not covered*
Africa	71
Asia and Pacific	56
Latin America and Caribbean	51
Total	59

Source: World Health Organization data in *World Water* (1986).

During the 1980s the actual coverage has undoubtedly improved. Under the aegis of the International Drinking Water Supply and Sanitation Decade, expanded efforts have been mounted to reduce the number unserved. Yet even under optimistic assumptions, the number unserved in 1987 is still about 50 percent. In some areas, progress is much slower than in others. In Africa, two of every three rural people still do not have access to an improved supply. In Asia and the Pacific, where 40 years ago coverage levels were similar to those in Africa, rapid progress has been made. In Latin America and the Caribbean, coverage has historically been higher than in other regions, but slow to rise further.

Continued population growth will make future efforts to increase coverage even more difficult, and will add to the problems of catching up with past neglect. Rural populations are still growing rapidly (Table 2), despite migration to urban centers. The situation is especially difficult in Sub-Saharan Africa. Many observers would consider it unduly optimistic to hope that the number of rural people with access to safe water might double by the end of the century. Yet, even if this is achieved, rural population growth will ensure that there are just as many rural Africans unserved in 2000 as there were in 1985.

Table 2. Rural Population Growth in Developing Countries, 1980 and 2000

Region	*Rural population (millions)* 1980	2000
Africa	330	450
Asia and Pacific	2,060	2,510
Latin America and Caribbean	120	130

Daunting as these challenges may be, they are not the end of the story. Most developing countries face severe financial, technical, and institutional constraints. In many cases, the easier, less expensive areas have been tackled first, leaving the harder tasks ahead. And, while new construction goes on, the problems of keeping existing systems functioning continue to grow. Although there are no global data, informed opinion has it that one in four rural water supplies is out in most developing countries, and that in some countries construction of new facilities is not even keeping pace with the failure of existing facilities.

In many cases, maintenance of the systems has depended heavily on the central government for finance and technical support. As more systems are built and commitments grow, pressures on government spending are becoming more acute. In short, disappointing as progress may have been, there are doubts both about the prospects of sustaining schemes that have been built and about extending services to those still unserved, unless a different approach is taken.

Conventional top-down approaches, with central governments and external agencies taking dominant roles, have missed the mark often. To mention two examples from many: in one Southeast Asian project, the government dug wells, installed handpumps, and committed itself to maintaining them, only to find that the people continued to use their traditional surface water sources; in an East African project, improvements were built without adequate planning to support the recurrent costs, with the result that even though the people welcomed the new facilities and wanted to use them, the systems fell into disrepair and disuse.

The problem is not, as once was thought, that the available technologies are too complex to be used by rural people. Certainly, some technologies are excessively complicated, but others exist now that are simple enough for anyone to operate and maintain, even with minimal training. Also, rural people have greater skill in handling new technologies than is sometimes appreciated. It is ironic but true that inoperative water supply systems are found in areas where bicycles, radios, irrigation pumps, ceiling fans, and small industrial machines are reliably maintained. It is apparent that what is at fault is not the complexity of the technology, but the top-down approaches being followed in the design of systems and the maintenance of facilities. Where people regard a water supply scheme not as their own undertaking but as something done for them by the government, they do not feel as responsible for the scheme's maintenance.

The amount of money that past approaches have required of governments and donors -- and would require in future if the goal of universal coverage is to be achieved -- is yet another difficulty. As noted earlier, the sum required to reach universal coverage, even assuming the minimum estimate, far outstrips any realistic projection of what will be available from government budgets and external funding sources. This issue has become particularly acute recently, as more countries have adopted major macroeconomic policy reforms aimed at getting their economies on a sounder footing. The reforms have been necessary responses to changing economic conditions and will have important long-term benefits. Yet in the near term they imply added stringency, further limiting the public funds available for sectors such as water supply and sanitation (Box 2).

Significant country-by-country differences also exist. In much of Asia, where economic polices have adjusted rapidly and effectively to changes in the world economy, rural development has proceeded, and, in many cases, sustainable and replicable rural water supply programs are well established. In Latin America, some countries have developed sound rural water supply models, but many others have not. In those that have not, progress will be slow.

In Sub-Saharan Africa, where financial and institutional limitations are particularly severe,

Box 2. Rural Water Supply Policies and Structural Adjustment

Most developing countries are now engaged in a process of structural adjustment aimed at restoring balanced, sustainable growth to economies suffering from severe distortions. Macroeconomic reform programs of the past few years reflect an increasing convergence of views about the types of policies that should be undertaken, such as restraining demand, reducing the size of the public sector, reforming institutions, liberalizing financial markets, and restructuring the trade regime. In this new policy environment it is essential that policymakers in each sector both assess the consequences of the macroeconomic changes for the sector and ensure that sectoral policies reflect broader development policies.

The structural adjustment process imposes two kinds of costs. Stabilization costs are incurred in bringing aggregate demand in line with aggregate supply. Restructuring costs are incurred in improving the efficiency of resource use through changes in the overall incentive system and relative prices and through shifts in government expenditures and revenues. While almost everyone suffers in the short term from stabilization, structural reforms produce both losers and winners in both the short and long terms.

The aggregate effect of adjustment on a particular sector is difficult to predict, in part because it depends on the conditions in each country. Box Table 2.1 outlines the main features of a typical adjustment package and suggests how each component may usually affect the rural water supply and sanitation sector.

Box Table 2.1. Typical Structural Adjustment Package

Feature of the structural adjustment process	*Likely effects on the rural water supply sector*
Stabilization	
Restrain public expenditure	Reduced government investment program and recurrent budget
Adjustment	
Correct bias against agriculture	Increased rural incomes, therefore increased income available for investment in improvement projects
Remove subsidies	Higher levels of cost recovery
Reform exchange rates	Increased costs of imported components (such as pumps, pipes, and energy) of projects
Reduce interference in financial markets	Higher interest rates, therefore increased annualized costs
Reassess the division of responsibilities between the private and public sectors	Greater role for private sector.

and where only a few countries have developed sound bases for expanding rural water supply services, the challenges are immense. In several of these countries, there have previously been crash programs to drastically increase rural water supply coverage. This experience, almost totally unsuccessful, provides a sharp contrast with that of countries (including neighboring ones) that have been less ambitious, but more patient and persistent in developing models for community water supplies.

The first step must be to develop appropriate policies for rural water supply, the second to develop locally relevant models for translating these policies into practice, and the third step that must be taken is to develop mechanisms for replicating successful small-scale experiences in a national program.

For many this time-consuming process is hard to accept, for there are desperate needs now. But the experience of the past 20 years has shown clearly that rural people would have been much better off had governments and external agencies pursued a patient, coherent approach and developed a sound basis for meeting long-term needs.

Advances in Low-Cost Technologies

Since the mid-1970s, significant efforts have been undertaken by the United Nations' Development Programme (UNDP), the World Bank, the United Nations' Children Fund (UNICEF), and numerous bilateral agencies and national governments to develop and improve low-cost water supply and sanitation technologies for use in developing countries.

Previously, the available handpumps had mainly been designed to meet the needs of individual families and were unable to withstand the heavy use typically given a community handpump in developing countries. New efforts have been made to improve the engineering designs of handpumps by experimenting with different operating principles, different designs of key components (such as pumprods, bearings, rising mains, cylinders, and seals), and the use of modern materials (such as plastics). In the 1970s UNICEF sponsored extensive research and development work in the Indian subcontinent, and made important advances in designing sturdy deep- and shallow-well handpumps.

More recently, the Handpumps Project which is funded by the UNDP and executed by the World Bank, has undertaken large-scale laboratory and field testing of 70 different types of handpumps and has rated these according to criteria such as performance characteristics, reliability, ease of manufacture and cost. The Handpumps Project has promoted the concept of VLOM (village level operation and maintenance) as a way to overcome some of the obstacles to establishing sustainable handpump systems. Some major advances in both hardware and software have been made, and other critical technical problems (such as developing VLOM handpumps for lifts over 25 meters, and reducing the costs of well drilling) are now being addressed.

At the same time, intensive efforts have been focused on developing low-cost sanitation technologies. National agencies in some countries (including India, Viet Nam and Zimbabwe) have made major developments, with global efforts spearheaded by a low-cost sanitation project funded by the UNDP and executed by the World Bank. It is now recognized that either pour-flush or ventilated improved pit latrines are appropriate technical solutions for the excreta disposal problems encountered in most rural settings. The focus of low-cost sanitation activities is now turning to the large-scale implementation of sanitation schemes that use these technologies.

The Experience of the World Bank

The World Bank began lending in the water sector in 1961, but did not begin activities in rural areas until the early 1970s, when modest rural water components were included in some agriculture and rural development projects. From 1974 to 1984, rural water and sanitation components represented less than 3 percent of the total costs of the projects of which they were a part. Through 1984, the Bank's first ten years of rural water supply and sanitation lending had amounted to a total of about US$500 million.

While past investment by the Bank in the sector have been relatively small, this is now changing. Considering all World Bank expenditures on water supply, the proportion of funds allocated to rural projects almost doubled from an average of 8 percent between 1974 and 1980 to an average of 14 percent from 1981 to 1985. The trend in Bank rural water lending is illustrated in Figure 1. Water supply components of overall rural development projects are still important, but there is a trend toward stand-alone rural water projects, including a US$80 million credit to China and a US$16 million loan to Brazil.

Figure 1a. Rural Water Supply Projects and Components in Water Supply and Sewerage Sector
(US$ Millions)

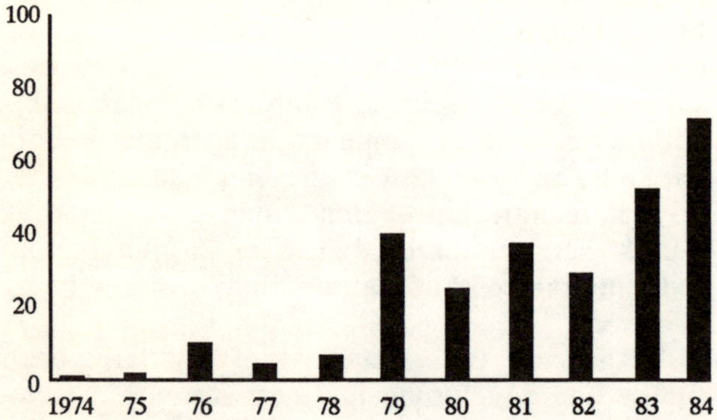

Figure 1b. Rural Water Supply Components in Agriculture and Rural Development Sector
(US$ Millions)

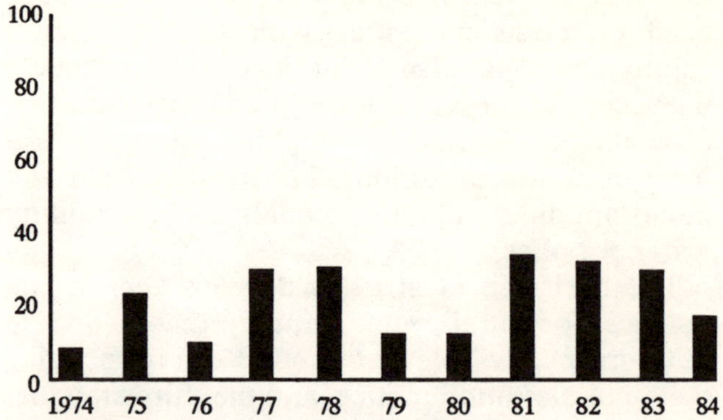

In the past, rural water supply activities funded by the Bank have faced difficulties similar to those faced by others working in the sector. This was partly because many of the rural water supply components were such small elements of agricultural and rural development projects that little attention was paid to their design. A 1985 Bank review of the sector concluded that overall performance was disappointing, and suggested that technology did not appear to be a major problem. The problems were diagnosed as being that the improved systems often did not meet local expectations, local institutional realities were not taken into account, and government support was erratic.

One positive result of Bank involvement in the sector has been that its insistence on the importance of certain concepts (such as cost recovery) has been instrumental in persuading governments and other donors to accept these principles. It is also clear that, although some problems persist, recent rural water supply projects' designs have applied many lessons. A central feature of the current China rural water supply credit is to develop provincial revolving funds, and an objective of the current Brazil rural water loan is to develop sound institutional and financial policies before large sums are invested in the sector.

2. What Needs to Be Done

The themes introduced in Chapter 1 stressed the role that users and governments must have:

- *Users* must decide on the type of improvements to be made.
- *Users* must pay most of the costs of the chosen services.
- *Users* must take responsibility for maintaining the facilities they have chosen and built.
- *Governments* and external agencies must establish the environment in which communities can construct, operate and manage improved facilities.

As this approach is implemented more widely, governments, donors, and communities will need to introduce substantial changes in what they do and how they do it. This chapter discusses in more detail the types of changes that will be necessary and why they are needed.

The main points are best introduced by examples of actual experience in the sector. The examples happen to be from East Africa and South

Box 3. The Lessons of Experience: Examples from East Africa

In more than two decades, a wide range of strategies for improving water supplies has been tried in East Africa; some have been successful and others have not. Important lessons that should be taken into account have emerged from this experience.

Kenya

In one large project begun in the 1960s, a top-down strategy was followed. Teams of experts linked to central authorities were to come into each village, decide what was needed, plan and construct the necessary facilities, acquire and install the required equipment, and make arrangements for centrally supported, regular maintenance. Access to the improved water supply was to be free or almost free to everyone in the village, on the philosophy that no charge should be collected for an essential, basic need such as water. Villagers also were not expected to commit significant resources to the effort in any other way; thus virtually all of the costs were to be covered by funds from outside the village. Practically the only role for the community was to receive and use the finished product. The project's outcome fell far short of expectations. In the end, most of the improvements were no longer operational. Large investments had brought little lasting result.

In other efforts in Kenya during the same period, a different strategy evolved, which proved to be more successful. Ironically, these efforts originated not from a water supply program itself, but from a combination of broader policies that helped to promote rural development overall. Government policies -- making good land available to small farmers and maintaining high prices for cash crops -- produced substantial gains in real rural incomes. In addition, the government encouraged the activities of private groups. One popular type of communal development activity was improving water supplies. Undertaken with virtually no technical assistance, these schemes cost much less than the government schemes, have been improved gradually over time by the communities, and are among the more reliable rural water systems in Kenya.

Tanzania

After independence, the central government adopted objectives that called for using government-provided rural water supply improvements as a way to increase the general welfare of the population. The government took full responsibility for building, running and financing rural water systems. In practice, however, it was unable, financially or institutionally, to maintain and operate the systems, which were primarily diesel-driven piped systems with extensive distribution networks. The legacy today is primarily one of abandoned, nonfunctioning systems.

Unlike in Kenya, community-based systems did not develop to fill the gap. One reason for this was that Tanzania's overall policies toward rural development were different from Kenya's. In Tanzania, rural production was heavily taxed and private sector initiatives were discouraged.

Malawi

Malawi has followed an approach very different from those of nearby Kenya and Tanzania, and, on the whole, the differences have led to more successful outcomes. Since independence, the government has strongly opposed the proposition that it should provide heavily subsidized services to rural people. Over time this position was reinforced because Malawi attracted relatively little aid from countries with a strong tradition of state subsidization of services. Instead, the government policy was that rural communities would develop primarily through self-help activities, supplemented where necessary, by technical and training support from government, and by materials from donors.

Starting in just one community of 2000 people in 1968, the Department of Community Development gradually developed a highly successful procedure for involving the community in the planning, construction, and operation and maintenance of these systems. As experience was acquired, the scope of the program was gradually expanded, to a point where currently nearly 1 million people have high-quality, reliable, and convenient water through schemes that are owned, built, and maintained (with critical but limited government support) by the communities they serve.

Initially the program was concentrated in areas appropriate for gravity-fed piped system, which are labor-intensive and thus ideally suited to self-help. The lessons of this experience are now being adapted to the different circumstances faced by those rural people who must be served through groundwater supplies and for groups of users in the fringes of small urban centers.

Box 4. The Lessons of Experience: An Example from Southeast Asia

The Northeast Thailand project, funded by the U.S. Agency for International Development, was initiated to improve the health of rural people by providing safe alternatives to contaminated water obtained from unprotected traditional sources. The northeast was a priority for government investment because it is one of the poorest areas in the country. So, when the first project was designed in the early 1960s, it was assumed that villagers could and would pay little for an improved supply. Accordingly, the target was to provide protected water at minimal cost. Since groundwater is abundant in the region, the technology chosen was handpumps.

Five years later the project was evaluated. Most of the handpumps were not working, and the people's water use habits were largely unchanged. Consistent with conventional assumptions, the failure was attributed to a technology that was too difficult for the villagers to maintain and the inability of poor villagers to pay for improved water supplies.

In a follow-up phase, motor pumps provided piped water at community standpipes. Again, the project failed. Five years after implementation, 50 percent of the systems were not working at all, and another 25 percent operated only intermittently. The problem was again initially put down to complex technology, weak institutions, and an inability to pay.

Gradually, however, it became apparent that the main problem was not the capabilities of the villagers, but the fact that the service being offered was not what they wanted. They did not want handpumps, which were not considered any significant improvement over the commonly used rope and bucket. Standpipes were no closer than their traditional sources and so offered no obvious benefits. Only water piped to yardtaps could meet the people's aspirations, as the time saved collecting water and the apparent high quality of the service were thought to be worth paying for.

Potential problems in providing this higher level of service were clear: the systems would be more complex and more difficult to manage and maintain, and the price to be paid for the water would be high – even more per liter than people paid in Bangkok. Project staff were surprised when villagers responded that they could and would pay the amounts required, that diesel fuel could be purchased and pumps maintained, and that trained people would run the systems if they were adequately paid and were supported by local government water officials.

The level of service was changed. Yardtaps were allowed, with the users paying the full costs of connection. Five years later, the verdict was in: 90 percent of the systems were functioning reliably; 80 percent of the people were served by yardtaps; large economic benefits were perceived, such as timesavings, gardening, and livestock raising; pumps, treatment works, and distribution systems were maintained; and locally adapted financing systems had been developed with meters installed and regular payments sufficient to cover operation and maintenance costs, major repairs and some depreciation. Not only had the systems been maintained, but because the service was so popular, many systems had extended distribution lines to previously unserved areas.

and East Asia, but their lessons reflect the experiences of many improvement projects throughout the developing world.

The first three examples are from Kenya, Tanzania, and Malawi (Box 3). In Kenya, one project tried a government-centered approach, while another, independent, initiative resulted in a community-centered approach. The government-centered strategy was unsuccessful. The new facilities, which had been designed by the government and constructed with little or no community involvement, were not adequately maintained, and most soon stopped operating. The project had sought to make safe water free to all, but the large subsidies were neither sustainable nor replicable. By contrast, the community-based strategy, which emerged more by chance than by design, has been effective and highly beneficial. In Tanzania, a government-centered approach failed when the government agencies responsible did not deliver the financial and technical support they were supposed to provide for maintenance and operations. In Malawi, a community-centered approach was highly successful and spread from one village of 2,000 people in 1968 to many villages covering nearly 1 million people in less than 20 years.

Some observers have concluded from such instances that simple, low-cost systems -- for example, a central handpump well with no piped distribution -- are always the best choice. In fact, although that may frequently be the case, it is not always so. The correct inference to draw from the evidence is slightly different -- as demonstrated by the fourth example, taken from Thailand (Box 4).

During a 20-year period, several different strategies were tried in a poor rural region of northeast Thailand. Systems delivering a low level of service (through handpumps or communal standpipes) did not succeed. Initially, it was thought that this was due to complex technology, weak institutions, and an inability to pay. But when villagers were offered the option of yardtaps, a level of service they perceived as being a real improvement over traditional sources, users paid large amounts, systems were maintained, financing systems were developed, and services were extended to previously unserved areas.

The final example is the national program in

Box 5. The Lessons of Experience: The Example of India

India has made remarkable progress during the Water Decade. Between 1980 and 1983, the percentage of the rural population reported to have access to a protected water supply increased from 31 to 47 percent. Impressive as this is, recent analyses have identified areas that need attention if what has been built is to be fully used and sustained, and if further gains are to be made.

The first problem is that, as in other countries, official figures of access probably give an overoptimistic picture of the actual situation. A survey by the government and UNICEF of handpumps in Orissa, Madhya Pradesh, Tamil Nadu, and Rajasthan showed that only about one-half of the villagers ostensibly served by handpumps were actually using them.

The second problem is the familiar one of operation and maintenance. In most of India, the water supply in rural areas is maintained by either state public health engineering departments or village councils (panchayats). The panchayats are unable, technically or financially, to maintain piped water systems, which thus function only where state governments take care of the maintenance. No institution is responsible for maintaining nonpiped supplies, although state governments do try to help. Only in rare cases do local bodies have a defined role. In such maintenance systems, the community does have one vital role -- to inform the authorities when systems are not functioning. However, the survey described above found that fully 50 percent of pump failures had not even been reported.

Although great strides have been made since the 1970s, when as many as 75 percent of handpumps installed in India were out of action at any given time, maintenance of rural water systems is still far from satisfactory. The recent survey found that, while new handpumps were usually functioning well, about 25 percent of the handpumps over five years old were not in working order, with more than 30 percent of these out of order for more than three months. The government estimates that inattention to maintenance is cutting the useful life of water supply systems in rural areas by 50 to 75 percent.

A major cause of the maintenance problem is the policies that have been pursued for financing rural water supplies. In rural areas, there are no charges for water, nor are any funds for water service recovered from the population. The Government Working Group on Operations and Maintenance has stressed the need for the beneficiaries to cover at least operations and maintenance costs; a point strongly endorsed by high-level central and state government policy makers.

It is now clearly understood by the government and donors alike that certain policies have to change. The government National Water Master Plan stresses that the local community has an important role to play -- including the contribution of money, land material, services, and labor. UNICEF has proposed that village water committees be established and that their functions include raising funds for preventive maintenance. The government Planning Commission has summed up the situation well:

> Lack of involvement of the local communities in the maintenance arrangements, shortage of staff and inadequate funds are the main reasons why existing water supply schemes have failed to yield the expected results. It is clear that the operation of the small rural water supply system can only be ensured with the participation of the village community. It has been noticed that wherever maintenance arrangements have been adequate the beneficiaries are not unwilling to pay nominal charge for the water supplied to them. The effort should be in all cases to recover at least the operating costs.

India. India has made major gains in meeting the rural water supply needs of its people, yet difficulties are being encountered. Box 5 describes steps being initiated by the central and state governments to strengthen the national program by redefining the roles and responsibilities of the government, the community, and the private sector and by increasing community involvement in financing and maintaining the improved water supply systems.

The lessons from these examples are important: people *do* want improved services, but only if these meet their perceived needs; the appropriate technology is not always the simplest and cheapest; users are able and willing to pay for services that they perceive to be valuable; consultation with the communities is essential in choosing technical and financial solutions; careful attention needs to be paid to the capacity of local and national institutions to operate and maintain the system; local institutional problems can be resolved if a desired service is offered; central governments should generally not try to operate and maintain rural water supply systems directly, but do have essential other roles to play; service can be extended to people who had been unserved, if the services can be sustained and are considered as valuable.

How then should governments, donors, and communities proceed in making the transition to policies and programs that are more in accord with these lessons? The most important steps involve answering a number of basic questions:

- What is it that the community really wants and is able and willing to pay for?
- What should the community take charge of?
- How much can it do? What support does it need?
- How much of the costs can be covered from within the community itself? By what

mechanisms (fees? a tax?) should it mobilize its financial and nonfinancial resources?

- As higher-level authorities reexamine their own role, what should be their stance toward the role of private sector suppliers of, for example, drilling equipment, and construction and maintenance services? In short, what should be the distribution of responsibilities between the public and the private sectors?
- What should higher-level authorities do to strengthen their own institutional capabilities to carry out their appropriate role effectively?
- Given that the role of women in community water supply is crucial and frequently misunderstood how should improvements take it into account?

Providing Services That People Want and Are Willing to Pay for

Experience has shown that when efforts to improve community water supply go awry, it is often because first things have not been put first. The first thing every new initiative must do -- the only good starting place -- is to determine what improvements in water supply the villagers want and are able and willing to pay for.

What Do Villagers Want?

The potential benefits of improved water supplies have already been stated. But what the residents of communities *perceive* the benefits to be is often a different issue. Promoters and users of improvement projects typically have different perceptions of the benefits of improvement. For governments and donors, the health issue is usually paramount. For the villagers themselves, other factors are often seen as more important, and the prospect of improved health is frequently viewed with scepticism. This is hardly surprising; as was mentioned earlier, such improvements are a necessary, but not usually sufficient, condition for health gains. Even well-designed scientific studies have found it difficult to isolate the direct effects of water and sanitation improvements on community health.

In most rural areas, the most important benefits perceived by users are convenience, resulting from the reduced time spent fetching water, and esthetic and economic benefits resulting from the availability of larger quantities of water in home. In many settings there is also the inherent attraction of "modern" service, similar to those available in towns and cities. Having direct piped service to the yard or even house is often a strong preference. In Malawi villagers are prepared to make significant payments in cash and kind for piped water, even where travel time to the water point is not reduced. In Zimbabwe willingness to pay for standpipe supplies is greater than for water from a handpump. Improved water supplies may also be valued because they reduce conflict at overcrowded sources, as studies in Guatemala, El Salvador, and Bangladesh have shown.

What people want in water supply cannot be understood without also considering what they want in sanitation services. Although this report is not about sanitation services, several points about sanitation are important in water sector planning. People throughout the world have an array of strongly held beliefs about defecation, but where improved facilities are wanted, surveys in India, Malaysia Mexico, Zimbabwe, and other parts of the developing world show the principal perceived benefits to be privacy, convenience, and status -- but not health. It follows that where population densities are relatively low, demand for improved facilities for excreta disposal is rarely a high priority.

Rural people almost always want improved water supplies, but seldom want to invest in sanitation improvements. A recent survey in Zimbabwe, for instance, showed that 26 percent of the women and 20 percent of the men saw improved water supply as their most important development need. Although there has been a relatively successful effort to improve excreta disposal facilities in rural areas of the country, less than 1 percent of the women and no men put improved excreta disposal at the top of the list. There is, however, consistent evidence from Asia, Africa, and Latin America that if improved excreta disposal facilities are promoted in conjunction with a water supply program that meets the felt needs of the people, improved latrines will often be built and used.

It may seem self-evident that village water supply and sanitation schemes should meet users' perceived needs. In many instances, however, the level of service and even restrictions on how the water is used are imposed by a central agency without considering the community's wishes. On two recent schemes in Kenya, for example, in part because of prohibitions on the use of water for nondomestic purposes, water payments were not made and the schemes were poorly maintained.

How Much Are Villagers Able to Pay?

Funding agencies have generally assumed that rural people are able to pay only very little for an improved water supply. The Northeast Thailand demonstrated the fallacy of this assumption, for villagers showed a considerable ability to pay in cash for the services they wanted. The same is true in even poorer parts of the developing world.

Mali is officially the fifth poorest country in the world, yet a survey in one of the country's least developed and most remote areas showed that rural dwellers have considerable cash income from nonagricultural activities. Between 40 percent and 50 percent of total income came from cash-earning activities, while high expenditures on durable goods and social activities (such as community projects, celebrations, and gifts) and loans to others was evidence of ability to pay cash for valued services. Similar surveys in other countries confirm that cash or barter equivalent is frequently available. There is visual evidence too: there are few villages in which consumer items such as radios and bicycles are not common. Links with urban economies are strong.

There are undoubtedly some rural areas where cash is simply not available, but even in these areas very poor households usually have some resources -- in particular their own time -- to contribute to constructing and maintaining service improvements. This contribution can often be important; water supply construction costs were reduced between 15 and 45 percent in Latin America and Africa when well-organized self-help labor was used. These contributions naturally have most effect when labor-intensive technologies, such as gravity-fed piped systems, can be used.

How Much Are Villagers Willing to Pay?

Experiences from many poor countries demonstrate that rural people are willing to make contributions in labor and cash to the construction and maintenance of water supplies where there is a clearly felt need and where the new service is perceived to be a marked improvement over existing sources.

Development banks endeavoring to incorporate willingness to pay considerations into the design of rural water projects frequently assume that rural people are willing to pay a fixed proportion (usually 3 to 5 percent) of their income for an improved water supply. This assumption has proved to be too simplistic. In some settings people are willing to pay more than 5 percent; in others they will not pay even 3 percent. A Bangladeshi villager with a river at his door is prepared to pay much less for a handpump than a Malian with a similar income who has to walk several kilometers to get water. Egyptian villagers who would not pay 1 percent of their income for a public standpipe supply are nevertheless willing to pay more than 3 percent for yardtaps. Studies on water demand have generally found that poor people are willing to pay a higher proportion of their incomes for improved supplies than are their richer neighbors.

Willingness to pay depends on many other factors besides income (Box 6). Family characteristics and the convenience, reliability and perceived quality of existing and proposed water supplies are dominant factors. Indeed, it must be concluded that the 3 percent assumption, like the "can't pay" assumption, has been based more on administrative expediency than fact.

Because willingness to pay varies by area, there is a need for rapid, inexpensive, yet valid methods for assessing willingness to pay. Work aimed at developing such methods is described later in this report.

Putting the Community in Charge

How realistic is the concept, basic to the proposed new approach, that local people should have a leading role in planning, constructing, financing, and managing rural water supply and sanitation projects? Again there are good and bad experiences to draw on for an answer.

There are two important dangers. First, it is easy to romanticize "the community," expecting villagers in developing countries to demonstrate a cohesion, capacity, and will that do not exist elsewhere in either developing or developed countries. Second, it is wrong to trivialize the notion of community participation using this as a euphemism for low-cost labor. Proper community participation, however, has enormous potential, and there is enough experience to define conditions under which local people can indeed be in charge of successful rural water supply projects.

The first requirement is the capacity and will of the government -- ideally at the local level -- to support the endeavors of local people. Boxes 7 and 8 give contrasting examples of government support. Box 7 tells what happens when projects are simply handed over to the community without appropriate preparation or support. The Malawi

Box 6. What Determines Willingness to Pay for Improved Water Services?

Consumers' willingness to pay for improved water supplies can be influenced by perceived benefits, income, characteristics of the existing source, and many other factors. Recent surveys on willingness to pay have added to data available from evaluations of some rural water supply projects and several general conclusions can now be drawn.

Perceived Benefits

Convenience, amenity, and economic benefits are important for users in surveys throughout the developing world, including Indonesia, Kenya, Lesotho, Peru, and Thailand. Water quality is important, too, but it is judged by taste, odor, color, and tradition -- not by bacteriological quality. In the Republic of Korea, for example, shallow well water was preferred to chlorinated piped water, and in Thailand many people, even those with private connections, would not drink chlorinated water. In Malawi and Brazil some households prefer water from traditional (contaminated) sources, because the tap water "smells of drugs."

Income

In Colombia, Kenya, Korea, and Thailand, surveys confirmed that families with higher incomes were more willing than poorer families to pay for connection to an improved water supply. In Chile, for every 10 percent increase in income, families consume 4 percent more water per capita. In Malawi, in areas where users have to pay for water from standpipes, the use of the standpipes increases following the sale of cash crops. Surveys in Zimbabwe and Haiti showed better-off families to be willing to pay 80 percent more for improvements than their poorer counterparts.

Water Charges

In Colombia, Indonesia and Panama, use of the improved service was higher when the price charged was lower. In rural Chile, a 10 percent increase in price was associated with a 3 percent drop in consumption.

Other Prices

In rural Indonesia, electricity is considered a higher priority than water connection and users are unwilling to pay more per month for water than they pay for electricity.

Value of Women's Time

In Korea, connection rates are higher for families where women are educated and employed because the value of their time is higher. In Kenya, pastoralists were not prepared to pay for improved domestic water supply schemes, in part because the male decisionmakers believed that there was no value to freeing time for women, because they did not help to care for cattle. For agro-pastoralists, however, women play a significant role in agricultural production, so people are much more willing to put money and effort into improved water schemes. The Zimbabwe survey showed that women were 40 percent more willing than men to pay for standpipe supplies.

Level of Service

In Thailand and Tunisia, users were not prepared to make relatively low payments for standpipes, but they did pay substantially more for house connections. In Egypt, villagers were prepared to pay only 1 percent of income for standpipes but 3 percent for house connections. In Indonesia, where flow restrictors are used, willingness to pay for house connections is greater when flow rates are greater. The recent surveys in Haiti and Zimbabwe revealed that willingness to pay for yardtaps was 1.3 times and 2.3 times higher, respectively, than willingness to pay for standpipes. In Zimbabwe, willingness to pay for standpipe supplies was, in turn, two to four times higher than willingness to pay for water from a handpump.

Characteristics of the Existing Source

Where traditional supplies are deemed acceptable (such as private wells in Korea and Mali, reliable springs in Peru, wells and springs in Indonesia, and surface water supplies in areas of Tunisia), willingness to pay for improved services is relatively low. In areas of Kenya where water is scarce, villagers organize and pay for operation and maintenance of boreholes, but in areas with surface water, people are reluctant to pay for water. In Ghana and Kenya, the number of people using handpumps drops by 50 percent or more during the wet season, when surface water is available. In the Haiti and Zimbabwe surveys, willingness to pay rose substantially as the distance to existing sources increased.

Other Productive Activities

Where water can be used to increase the home production of crops and small animals, willingness to pay is likely to be higher. In Kenya, for instance, those with home gardens are willing to pay for improved domestic water supplies, while those who raise cattle are not.

Credibility of External Agency

Experience in Colombia, Malawi, and elsewhere shows that willingness to pay is relatively high when agencies have proved that they can deliver what people want. In El Salvador, Guatemala, Haiti, and Lesotho, however, people were less willing to collaborate and to pay because of bad past experiences nvolving abortive attempts to introduce innovations, compulsory programs or activities by a politicized agency.

Box 7. Community Management with Inadequate Government Support

About 20 years ago, Peru's Ministry of Health began to install potable water supply facilities in coastal and highland areas. The ministry went beyond installation, however, and attempted also to stimulate local community participation in approving, financing and managing the water systems.

The "promotion" or preconstruction phase is generally marked by enthusiasm (especially from women), cooperation, and anticipation. Enthusiasm persists into the second phase -- construction -- in which the water program personnel concentrate on technical matters, and villagers provide labor and elect and organize a Water Board. The second phase generally culminates on a euphoric note, with the inauguration of the new facility.

Having installed the system and paid most of the costs, the Ministry of Health retains title to it, but responsibility for the operation, upkeep, and general administration is turned over to the community's Water Board. This third phase differs profoundly from the previous two. Villagers find themselves inexperienced and untrained for their management role. Human and material resources are inadequate to cope with technical malfunctions. As a result, the Water Board and sections of the community become embroiled in controversy over water management issues.

Frequently either the Water Board or a faction at odds with the board appeals to regional water program officers for assistance in one or more of three kinds of problems: technical, financial, or administrative. Water program personnel usually respond quickly and effectively with technical assistance and can often either help or advise on financial problems. But when management difficulties arise, they are presented with problems outside their sphere of competence. Expertise in hardware and the technical aspects of rural water supply does not equip the officers to deal effectively with the many human issues involved in village water management.

A recent evaluation of Peru's rural water programs concluded that, while community participation is necessary, too little attention has been paid to a corollary issue -- community support. There is an equally important need for water program personnel trained in administration, management, and conflict resolution, to participate with, guide, and direct

and Colombia rural water programs described in Box 8 demonstrate how success comes when five critical conditions are fulfilled:

- Communities are involved in all stages of their water projects.
- The roles and responsibilities of the community and the government are clearly defined, and obligations are fulfilled.
- The government agency acts as a supporter of the community, not as owner and manager of the water supply.
- Contact between the communities and the government agency is through cadres whose primary skills are organizing and motivating communities.
- Government agencies fulfill their limited but vital tasks of motivation, training, and technical assistance.

A related requirement is a realistic assessment of the capabilities of both the community and the government. It is important that the community should not take on responsibilities that it does not have the resources to execute, and equally important that government should not try to do what local people can do better. In both Colombia and Malawi, local people assumed primary responsibility for construction and for operation and maintenance tasks, forming local institutions (committees) with prescribed mandates for this purpose.

While it is useful to talk about "the community", it is important to realize that some community members are more interested than others in improving water supplies. In rural water supply, probably more than in any other sector, sustainability is dramatically enhanced when women have key responsibilities. As the community members who literally shoulder the burden of fetching and carrying water, they have the most interest in seeing improvements made and sustained. That interest has translated into effective programs in many countries when women have taken on management responsibilities. This point is discussed in more detail later in this chapter.

Some key services, such as drilling of wells, construction of sophisticated water-retaining structures, and servicing of machinery, are best performed neither by government nor by the community, but by the private sector. This intermediate role for the private sector can be the difference between success and failure. A properly equipped and suitably rewarded private sector can be an effective bridge between the government's limited capability to service dispersed communities and the community's shortage of skills, tools, and materials.

Institutional arrangements for productive partnerships between the people and the government vary widely. Even in a confined and relatively homogeneous area of Mexico, for instance, great variety has been found in local arrangements for maintaining, administering, and financing water

Box 8. Community Management with Adequate Government Support

Colombia

Colombia has deservedly earned a reputation for having the best rural water supply program in Latin America. The program is a model of effective cooperation between villagers and government. From sporadic efforts to construct seven rural water projects in 1963, the program brought access to safe water to 80 percent of the country's rural population by 1980.

Over the years, the National Institute of Public Health (INS) developed a standard and detailed methodology for working with villagers to improve their water supplies. In each phase -- community study, project preparation, community organization, and motivation, construction, and administration and supervision -- the responsibilities of the INS and the community are clearly spelled out.

Community Responsibilities

The local beneficiaries play a major role in all stages, from initiation to operation and maintenance. Such tasks include:

Study: make suggestions on the basis of community experience; collect the required socio-economic and environmental information.

Project preparation: participate in the topographic survey; provide the necessary local data; elect the Administrative Committee.

Promotion: contribute expertise on community organization; participate in project meetings.

Construction: provide materials, labor, cash and transport; carry out social functions to raise money.

Administration and supervision: contribute financially and organizationally to the Administrative Committee, which is responsible for operation, maintenance and administration; ensure that system regulations are obeyed; suggest what extensions should be made to the system.

Government Responsibilities

The central office of the INS is responsible for developing design standards and instruction materials, planning, programming, and supervising activities. Close liaison with the communities is maintained through 24 decentralized sectional offices.

Promoters play a key role during all stages of construction, operation, and maintenance. They are trained in public relations, community development, local culture and customs, health education, and the basics of rural water supply and sanitation. The promoter helps to carry out a socioeconomic study before any promotion begins at the village level. He then helps to organize the community to set up the Administrative Committee, which includes members of local authorities and village development committees as well as locally elected leaders. As secretary of the Administrative Committee, which meets at least once a month, the promoter participates by auditing the books, helping to prepare monthly reports, helping householders who are in arrears with their water bills, and maintaining continuous contact with the assigned water supply projects. If operation and maintenance problems are beyond the capability of the local operator, the promoter obtains assistance from the sectional office, and when necessary pumps and motors are sent to commercial firms for repair.

The major shortcoming of the present structure in Colombia is the area of financing, a role for which a technical agency such as the INS is not well suited. In the future, it is likely that the INS will retain its technical role, but that the role of financial intermediary will be assumed by the National Housing Bank.

Malawi

Despite the quite different levels of development in the two countries, Malawi's rural water strategy bears a striking resemblance to that of Colombia. This suggests that general conclusions on the nature of effective community-government collaboration can be drawn from these two examples.

The Malawi Rural Piped Water Project is rightly regarded as the outstanding rural water project in Africa. Starting in 1968 from a modest cooperation between a single village of 2,000 people and the country's Department of Community Development, the project has since brought piped water services to nearly a million people. As in Colombia, the roles

schemes. The spirit generated by successful community-government partnership is illustrated by a remark by a government employee on the Malawi rural piped water project. Asked how he could possibly check on the activities of the widely scattered monitoring assistants under his command, he replied: "You do not understand. This is not the government's project but the people's project. The people know what the monitoring assistant is to do, and if he does not do it they report immediately to me. But this is not a problem we face very often."

People's commitments to manage and pay for water supply improvements are essential for sustainable and replicable investments, but the support role of government is also important. Where that support is inadequate, progress is impeded and problems occur.

Covering the Costs from within the Community

Earlier it was said that users should cover as much as possible of the cost of water supply improvements, since the alternative of relying mainly on funds from outside the community is often unsustainable and can lead to projects inconsistent with choices the users themselves would make. So communities need to think about cost recovery, of community and government are well defined:

Community Responsibilities

Project initiation: petition the government for a rural piped water project.
Study: assist in identifying water sources and conducting hydrological and topographic surveys; conduct meetings with Water Department staff to discuss community and government inputs and responsibilities.
Design: participate in decisions on the location of taps.
Construction: elect a main committee; provide housing for Water Department workers; organize digging.
Operation and maintenance: Elect repair and tap team members; raise funds for replacement parts; support government monitoring assistants; enforce community rules on water use.

Government Responsibilities

The Malawi project has always regarded the central skills of core government staff (engineers, field supervisors, and monitors) to be community organization and motivation, supplemented by the necessary technical skills. Responsibilities of government include:

Sector: raise necessary external funds; give training; set standards.
Study: make hydrological and topographic surveys; assess community organizational capacity; meet with community leaders to discuss community and government inputs and commitments.
Design: do the engineering design of the system.
Construction: provide materials and logistic support; supervise community digging and pipelaying activities; supervise specialized work done by private contractors.
Operation and maintenance: monitor system performance; supervise village repair and tap team members; provide technical services for tasks the villagers are not trained for, such as repair of asbestos-cement pipes; train community teams periodically in technical and organizational skills.

but what precisely should their contribution be? In addressing that question, it is important, first, to be aware of the broader arguments for cost recovery, which are summarized briefly in Box 9.

Until recently, cost recovery policies in the water supply sector have been dominated by the outdated view that rural people cannot afford to pay for the service. A 1970 survey of cost recovery policies in 122 developing countries (Table 3) showed that more than a quarter charged nothing for rural water supplies. Well under half of the countries aimed to cover even operation and maintenance, and in only very few was it policy to recover capital and recurrent costs. Since revenue collection is known to be weak in rural areas, actual levels of cost recovery were undoubtedly even worse than this survey revealed.

Table 3. Cost Recovery Policies in the Rural Water Supply Sector of 122 Developing Countries

Cost covered by user contribution	*Percentage of countries surveyed*
All costs (O & M plus capital)	6
O & M costs plus part of capital costs	16
O & M costs only	20
Part of O & M costs	30
No cost recovery	28

Note: O & M refers to operation and maintenance costs.
Source: Saunders and Warford (1976).

These figures are 17 years old. Since then, although some governments and funding agencies have revised their policies to achieve greater cost recovery, the overall picture has probably not changed significantly.

How much of the costs should a community aim to recover? From what has been said earlier and in Box 9, it follows that communities should be moving toward an eventual goal of covering all of the costs from their own resources. In some communities, substantially increased cost recovery is currently possible and appropriate, and it should be implemented without delay. In others, increases will need to be introduced more slowly, during a transition period that allows various political and economic constraints to be dealt with first. In communities where resources are most limited, a long-term view must be adopted that allows for continued financial support for the very poor in the near term, provided that the level of support is low enough to be sustainable and to not undermine local self-help incentives. This flexibility in the transition time must not be abused. The evidence cited earlier suggests that most communities can contribute more, in kind and usually in cash, than they do today, sometimes much more. The challenge is to strike a reasonable balance between moving too slowly or too quickly toward higher cost recovery.

Full recovery of operating and maintenance (O & M) has often been proposed as a practical middle course, leaving capital costs to be met by sources outside the community. This option has both advantages and disadvantages. Its main

Box 9. Why Cost Recovery Is Important

The reasons why the costs of water systems should be recovered from the users are broader and more fundamental than opponents of this philosophy have sometimes realized.

One reason typically advanced for cost recovery is that if the local people themselves do not cover their system's costs, it is unlikely that anyone else -- governments or donors -- will be able to do so on an adequate, long-term basis in most communities. This observation is certainly true and, as a practical consideration, is a sufficient argument by itself. Yet there are other, more far-reaching reasons as well, which have to do with efficiency and equity in the utilization of resources.

When users contribute little or nothing to covering the costs of a service, experience has shown that there is substantial inefficiency -- waste -- in how resources are allocated and used. The symptoms are familiar: the providers of the service are unresponsive to users' preferences, facilities are in frequent disrepair because maintenance is weak, capital expenditures are not made when they should be, workers are in no hurry to get things done, the service is not reliable, and its hours of availability are limited. Providers are continually underfunded; not being dependent financially on user patronage, they have little to fear from failing to make improvements. The result is not just that service is of poor quality; it is also that the resources available to the sector are being poorly utilized. Investments are not yielding their potential returns. Scarce resources that could be of benefit in other sectors are being wasted.

This efficiency concern also has another dimension. When costs are not recovered from the users, they are typically financed from government revenues through taxes of one kind or another, since all costs have to be shown to be met somehow. These taxes adversely affect efficiency. They divert resources from productive uses, distort incentives, and involve higher collection costs. An opportunity is missed to draw on a large reservoir of potential revenue in a nondistortionary way, that reservoir being the amount that users would be willing to pay if called upon to do so.

There is an equity argument, too, for cost recovery. When users contribute little or nothing, a part of the population, often a large part, typically remains unserved or underserved because whoever is responsible for the service has insufficient resources to extend and improve systems. Thus, a fortunate few -- predominantly middle- and upper-income households -- receive subsidized services, while the rest -- mostly the poor -- get little or no service at all. A few examples are presented in Box Table 9.1.

When the effects of the taxes used to finance the subsidies are taken into account, the result is that the rich gain and the poor lose, with the poor in effect subsidizing the rich. Frequently, this also means that urban dwellers gain and rural residents lose.

Box Table 9.1. Who Benefits from Subsidized Rural Water Services?

Country	*Income Level*		
	Low	*Medium*	*High*
India			
Number of water points per locality	0.8	1.9	2.5
Kenya			
Percentage of families with connections	34	54	66
Korea, Rep. of			
Percentage of families with connections	53	66	77
Colombia			
Percentage of families with connections	16	18	24

Note: The definitions of income strata are different for each of the cases cited.

advantage is that recovery of O & M costs greatly improves the prospects that the system can be kept functioning. If the community cannot even cover its O & M costs, it is dependent on external funds, which typically have not been reliably forthcoming. Another advantage is that O & M costs are easier to recover in practice than are capital costs, considering the recovery mechanisms possible in rural areas and acceptance by community members.

The chief disadvantage is that communities may tend to stop at recovering O & M costs, when they should be going on, at some point, to cover capital costs as well. Choosing O & M costs as a target has no basis in economic principles; it is merely a pragmatic, interim compromise. If the compromise becomes a permanent policy, incentives will be adversely affected. Funding sources responsible for capital costs -- in particular government agencies covering the costs that the communities do not -- will have incentives to choose investments that minimize future capital costs, which usually implies higher O & M costs. Communities will have incentives to minimize O & M costs and shift more of the cost burden to capital investments. The net result is likely to be a pattern of investment and resource use that would be less efficient from an overall national perspective.

On balance, recovery of O & M costs only is a realistic near-term objective for communities that currently recover less, but it should not be the ultimate goal. Conceptually, it is far better to use total costs as the basis, and then, if it is necessary for some communities to adopt interim targets that are short of full cost recovery, to select those targets according to each community's capability.

Box 10. Cost Recovery in China

China's policy of self-reliance has resulted in villages being largely responsible for sector development and for providing their own water supply. Typically, a village is expected to pay 90 to 100 percent of the capital cost of the system and the full cost of operating it.

The community plays the leadership role in initiating the water project and in selecting the source, system technologies, and service levels. Plans for waterworks are submitted for approval to the county, which helps in the design and purchase of materials and decides on the need for subsidies. The provincial government in turn compiles plans submitted by the counties. In the past, the provincial government has provided only limited financing, but recently some provinces have helped to set up revolving funds for financing rural water projects. The central government has only recently become active, offering guidance and coordination and, in exceptional cases, providing financing assistance.

Most waterworks in rural areas are owned, managed, and operated by the villages with some collectively owned water companies formed in the more industrialized regions. The virtual absence of central government financing for rural water supply distinguishes China from most other developing countries. Local governments and entities are expected to provide finance for capital costs, and consumers almost always pay the recurrent costs. In villages, householders share the costs with funds derived from communal activities. Water charges normally cover the costs of operation and maintenance and occasionally some depreciation.

For village waterworks systems, the village accountant, working part-time, keeps records of receipts, payments, and inventories for the waterworks entity. For larger systems serving several villages, accounts are kept at the waterworks offices by full-time accountants.

There is a strong sense of pride and ownership, which is demonstrated by a willingness to pay the bulk of the cost in advance, and the fact that systems are almost invariably well operated and maintained. Thus, water supply investments in China are both sustainable and replicable.

Some communities are already able to recover a substantial part of their total costs (Box 10).

Cost recovery policies should also be consistent with the principle of marginal cost pricing: the amounts that users pay, whether in financial or nonfinancial contributions, should be as close as possible to the marginal cost of providing the water (that is, the cost of the next unit of water produced, as distinct from the average cost for all units). Although there are economies of scale for many components of water systems, because the development of new water sources usually involves increasing costs, the long-run marginal cost is usually greater than the average cost. Thus, marginal cost pricing usually will not be in conflict with assuring that costs are adequately covered. In practice, first priority should go to working toward cost recovery, since that will help correct the most serious problems. Once progress has been made in that area, refinements can be considered later.

A variety of different fiscal instruments can be used to recover costs. One possible instrument is a tax of some kind levied by the community on all residents. Alternatively, a tax can be collected on some other basis, for instance, from all landholders, all shopkeepers, or all sellers of goods in village markets. Such taxes are sometimes called by a different name, such as "water fee," but they are still essentially a tax in that they do not distinguish users of water facilities from nonusers or heavy users from light users. A further possibility, also implicitly a kind of tax, is to get residents to contribute labor and materials.

Instruments that distinguish users from nonusers are more difficult to implement effectively in many rural situations, but experience has shown that they are often feasible. Where village institutions are strong and there are legal sanctions against nonpayers, revenues can be collected for any type of supply. For instance, in the fringe areas of towns in the Philippines users' groups form standpipe committees. The standpipes are metered and the users' groups collect fees from each participating family to pay the bill. In Côte d'Ivoire, groups of handpump users collect revenue from users and manage the funds to maintain their water supply. In Burkina Faso and Benin, concesionaires successfuly operate public taps.

In the many rural areas where structures or institutions are weak, however, it has usually proved difficult to collect revenues from users of a public tap and even more difficult to get handpump users to make payments. Where water is piped to individual houses, revenue collection is greatly facilitated since the threat of cutting off service can be invoked. In most developing countries, flat rate charges are levied on house connections. In some rural areas of some countries, including Brazil, Chile, China, India, Malaysia, and Thailand, meters are installed and volumetric tariffs are in place.

For the very poor, additional considerations must be taken into account. Ideally, if a country or community wants to provide some sort of subsidy to poorer residents, it is better (for many reasons, but especially for efficiency) to do so through a form of general support rather than through special provisions in a particular sector. In cases where that approach is not adopted for political or other reasons and where special provisions for the poor are considered in cost recovery schemes, several options are possible. The concept of a "lifeline" service, applied initially for services in urban settings, can suggest useful approaches for rural situations as well. In the urban context, water policies with a lifeline feature enable people to obtain some minimal initial level of consumption (commonly about 20 liters per capita per day) for a very small fee, with consumption beyond that level charged at higher rates. Since it is primarily the poor who consume at the minimal level, the aim of providing subsidized services to the poor but not to the better-off is accomplished. For rural systems, access to one basic supply point (for instance, a centrally located standpipe) can be made available at a low rate, while access to more convenient supply points (such as yardtaps) can be at higher rates. Alternatively, if the community uses some form of tax, it can be levied in such a way as to fall more lightly on the poor than on the better-off.

It has been argued by some that certain special characteristics of water supply activities need to be considered as well in cost recovery policies -- characteristics that go beyond the points already mentioned here. For example, it has been said that the externalities associated with the potential health effects must be considered and that other market failures may also merit attention. As theoretical possibilities, these propositions are correct, but their practical significance is dubious. The evidence on externalities and other market failures is limited and not convincing. Even if they are important factors, it is questionable whether any adjustments on their account are really necessary or feasible for cost recovery policies that are based on the principles recommended above. This is especially so when provisions are made for the poorer residents in a community and for near-term constraints in the ability and willingness of the community as a whole to cover their costs. Perhaps, when other more pressing priorities have been dealt with, such refinements can be made; but not now, as a first priority.

There are interesting contrasts between cost recovery for water and cost recovery for sanitation services. The situation for sanitation services is much more problematic, because willingness to pay for such services is generally low, except in high-income groups and densely populated areas. The same arguments of equity and externalities have often led to rural sanitation programs being heavily subsidized. In most cases, however, coverage levels have been too low to achieve significant health gains, and the subsidies (as usual) have gone to the rich, not the poor. In Zimbabwe, for instance, only 16 percent of families with less than average income have benefited from subsidized latrine programs, while 55 percent of the richest decile of families have taken advantage of the subsidies available. There is some evidence from Malawi and Thailand that water supply improvements may be a necessary first step in stimulating demand for better sanitation and that focused hygiene education programs in conjunction with water supply projects may reinforce this effect. It follows that modest subsidized efforts to provide hygiene education should be part of rural water supply programs.

Remaking the Partnership between the Public and Private Sectors

In the past, it was typically assumed in most developing countries that rural water supply should be the government's responsibility. Central government agencies took on a large role for themselves, frequently ignoring the potential of private sector institutions and local initiatives. International aid agencies initially welcomed a centralized approach, as this appeared to be the best way to reach the largest number of people with improved services. It was argued that water systems have large economies of scale that can be realized only by having a single large provider. It was assumed too that rural communities had neither the resources nor the capacity to supply the necessary construction and maintenance services. It followed that these services were usually provided by the central government.

It is now evident that, in the rural sector, governments usually supply the service at a high cost and are responsible for considerable diseconomies of scale in both constructing and maintaining water supply systems. In Kenya, for instance, salaries account for 70 percent of the cost of operating government-run water supply schemes, while in Côte d'Ivoire, where a private company operates most water systems, only 35 percent of expenditure goes to salaries. Similarly,

Box 11. Cooperation between the Public and Private Sectors in the Côte d'Ivoire

Until 1967, the water company in Côte d'Ivoire, SODECI (Societe des Eaux de Côte d'Ivoire) was a wholly owned subsidiary of the French firm SAUR. Since 1967, SAUR has retained 47 percent interest, with the remaining 53 percent ownership held in Côte d'Ivoire (49 percent private, 4 percent government).

SODECI actually began operations in 1960, under a 30-year concession contract, in which the concessionaire finances, constructs, and operates water supplies. The contract was converted in 1973 to the *affermage* model, under which the public authority handles planning and construction and contracts out operation and maintenance, collection of charges, and consumer relations to the *fermier*, or private operator. The fermier is compensated only by water sales, and the contract sets the sale price. The price includes a surcharge collected by the fermier and paid to the authority to amortize initial investments. SODECI's fee is about one-third of the water tariff.

Also in 1973, SODECI's responsibility was expanded from Abidjan to the towns and villages throughout the country. By 1984, SODECI operated piped water supplies to 132 towns and 95 villages, and wells for a further 7,900 villages. In the past ten years, the proportion of rural people reported to have access to safe potable water has increased from about 20 to more than 80 percent.

The *affermage* system has served communities with piped supplies well. Despite rapid expansion, water supply in Côte d'Ivoire offers one of the highest standards in West Africa. Systems are well designed, equipped, operated, and maintained. Quality and pressure are uniformly good, consumption is metered, and water losses are low.

Several factors contribute to the good performance.

- By setting tariffs to cover costs in full, Côte d'Ivoire can finance its existing services. Water rates are among the highest in Africa, which means that consumers rather than taxpayers pay for the service. Rates for small quantities are low, which helps the poor to afford the water.
- During periodic tariff reviews, the government can carefully scrutinize SODECI's costs.
- SODECI is free (within contracted limits) to hire, fire, and compensate its staff. This freedom, plus a strong emphasis on training (SODECI has its own training center), enables the company to attract, train, and keep qualified people.

Performance has been less satisfactory in the maintenance of the widely scattered boreholes. As part of an overall sector reorganization, SODECI will assume full responsibility for areas with piped water systems, while operation and maintenance of handpumps will be progressively transferred to the villagers themselves.

in Pakistan, private contractors drill wells at rates considerably lower than those of the government.

Innovative national programs increasingly seek ways to use the capacities of the private sector to reduce construction costs and to improve the standard of maintenance. In Bangladesh and India, building contactors are now widely used to construct government-funded water supply systems; in the Dominican Republic, a combination of self-help labor and private contractors is used.

The private sector (including community organizations) is also used successfully in operation and maintenance. A private company in the Côte d'Ivoire, SODECI (Box 11), has earned praise for its upkeep of rapidly expanding piped water systems. Kenyan water systems built and run by private associations completely independent of the national water ministry are among the country's most reliable systems. In China, the community has the lead role in initiating projects; in selecting water sources, system technologies, and service levels; and in operating and maintaining rural water systems. And cooperatives are providing effective management services in Argentina, Chile, the Philippines, and parts of the Middle East.

Successful projects clearly demonstrate that the private sector performs better than government in a number of important roles:

- Providing technical support in design and construction
- Supplying specific technical services (such as well drilling)
- Supplying materials such as pipes and pumps
- Contracting for specialized construction tasks
- Fulfilling specialized maintenance tasks.

During the past few years, rural water projects in Bangladesh, Belize, Bolivia, Burkina Faso, Egypt, Honduras, Lesotho, Mali, and the Philippines have been designed so that maintenance is provided by local private sector mechanics. Even the best known government maintenance system -- the three-tier system in India -- has run into such logistic problems that Maharashtra, Rajasthan, and several other states have decided to abandon the system and to promote the development of private maintenance capabilities.

Another potentially important role for the private sector in several countries is that of financial intermediary. Agricultural credit banks, and the other finance institutions with offices in rural centers can readily help communities to organize charging systems and to operate revolving funds.

In many countries, particularly in Sub-Saharan Africa, private sector firms that might provide technical and financial services are weak or nonexistent. In part this is because government involvement has discouraged the entry of private firms into these markets. Accordingly, an important role for government and external support agencies is to foster the development of private sector capabilities in these countries. India provides a good example of what can be done: the India Mark II handpump, which was developed by UNICEF, has been manufactured widely by the private sector.

In many developing countries, community-based nongovernmental organizations (NGOs) constitute an important resource for developing water supplies that respond to local needs and capacities. Because of their familiarity with the local residents and their knowledge of the community, NGOs can often play a valuable role. This includes identifying improvements that community members want and have the capacity to build and maintain and working with communities to construct, operate, and maintain the facilities.

Redirecting the Role of Government

Just as the role of the private sector and the community must be redefined, so too, careful consideration needs to be given to redirecting the government's role in this sector. It is evident that in most countries governments have done both too much and too little: too much in that the government's responsibility has included many tasks that should be carried out by communities themselves or the private sector; too little in that the government has not performed adequately those tasks that governments alone can and should execute. What then are the things government can and should do in this sector?

The first requirement is a clear delineation of the responsibilities of the government, the community and private sector. The detailed configuration of responsibilities must be individually defined in each particular country, but the roles may be generally described as follows:

- *Public sector:* promoter, educator, regulator and, in some cases, financier or financial intermediary.
- *Community:* owner, builder, manager and operator.
- *Private sector:* provider of special skills, materials, and services to the community.

As "promoter and educator," the government has several important responsibilities. The selection of an appropriate source for a community water supply and the management of conflicts (between, for instance, agricultural and domestic uses of water) require adequate hydrological and hydrogeological information. In Mozambique, as an example, after independence in 1975, rational planning of water supplies could not proceed because of the paucity of hydrogeological information in many areas of the country. Accordingly, the government gave high priority to developing an adequate data base. In this, and indeed all countries, the government has the major role to play in collecting such information and in making it available in a form that is readily interpretable by community members and technicians.

The government also has a responsibility to spread information (through schools, health facilities, training and visiting programs, and radio and television) on the importance for health of safe water and improved hygiene practices. It has to help communities to identify and overcome the informational, technical, and managerial constraints that limit the capacity of the community to improve its water supply conditions. In addition, the government has a vital role in training professionals, technicians, and community members in the technical, managerial, and financial skills needed to construct, operate, and manage improved water supply facilities.

As "regulator," too, the government has a major role to play in the sector. First, in many settings, the use of water by one group affects the quantity and quality of the water resources available to other groups. In Bangladesh, for example, unregulated motorized pumping for irrigation is thought to be a major factor in lowering the groundwater table, thereby rendering many shallow wells fitted with suction pumps inoperable. Similarly, in Tunisia and other water-scarce countries, there are increasing conflicts over riparian rights and allocation of water to competing agricultural, industrial, and domestic uses. The government has the responsibility of identifying potential conflicts and adjudicating such disputes. In particular, it is the government that needs to ensure that the relatively small quantities of water needed for domestic purposes are taken into account in overall allocation processes.

Second, the government has a major role to play in protecting water resources from contamination. This will frequently involve the proclamation and enforcement of land use ordinances designed to protect watersheds. Third, the government has a responsibility to develop realistic standards for water supply equipment and for water quality and to ensure that these standards are followed when approving funding for community improvement efforts and when approving equipment and services provided by the private sector.

It is important to emphasize that, in the past, water quality and other service standards have been adopted unchanged from industrialized countries and have either been essentially ignored (especially in rural areas) or have actually been a brake on sector development. The recent World Health Organization Guidelines for Drinking Water Quality represent a significant and appropriate change in the concept of water quality standards. As pointed out by the WHO, "standards and regulations achieve nothing unless they can be implemented and enforced... and adoption of too-stringent drinking water standards could limit the availability of water supplies that meet those standards." Appropriate adaptation of the WHO guidelines will mean that, as regulator, the government's role will no longer be to erect obstacles to all less-than-perfect improvements, but to encourage the improvements that can be made.

A critical constraint facing communities in improving rural water supply conditions is the difficulty of obtaining loans to finance such works. In some cases, commercial banks, savings associations, or development banks will be appropriate institutions to act as financial intermediaries; in others it will be necessary for the government itself to act as the financial intermediary. For example, in China, a World Bank loan has been used to set up provincial revolving funds, which are used to provide loans to villages for water supply improvements. It should be stressed, however, that the skills required for this task are usually not found in the agency with primary responsibility for rural water supplies. In Colombia, for example, the Ministry of Health has been unable to manage the role of financial intermediary, a role which will now be assumed by the National Housing Bank.

Finally, in some countries where the private sector is not yet sufficiently developed to provide some of the services necessary for sector development (such as drilling, purchasing, and procurement services), the government may provide such services temporarily. The philosophy behind the assumption of this responsibility by government, however, should be that such services would ultimately be provided by the private sector, for the reasons already outlined. Every effort should still be made to encourage the private sector to provide such services and to reduce the government's role in this area as rapidly as possible.

For the government rural water supply authority, the role of "partner" introduces new and different challenges. Managerial, organizational, and motivational capacities become more important, while engineering skills are less crucial. Purely technical organizations (such as public works authorities in many countries) are not well suited for this role. Most success has been achieved through agencies specifically created with organizational and motivational skills in mind. Examples include: the Department of Community Development, which sponsored the successful Malawi rural water program; the Population and Community Development Association, which has developed a successful rainwater roof catchment program in Thailand; and the special promotion services within Colombia's Ministry of Health.

Modest reforms will seldom be enough to accomplish the necessary changes, and change will rarely be easy in the face of the endemic problems: inadequate financial and human resources, poor management, overcentralization and inadequate incentive structures. Given an improved policy framework, however, it should be possible to identify the public sector authority with the appropriate orientation, and to help this authority evolve a workable structure for rural water supply activities.

Reorienting Perceptions about the Role of Women

It is little wonder that when women are asked about their development needs, improved water supply frequently tops the list. Rural women, who are often simultaneously farmers, food processors, cooks, and mothers, work extraordinarily long and hard days. In many settings, life is made even more difficult because a substantial amount of time is spent fetching water. To take but a few of many documented African examples: in the lowlands of Lesotho, 30 percent of families spend more than two and a half hours a day collecting water while in East Nigeria, water collection can take up to five hours a day. Such findings are not confined to Africa. Similar situations can be found in Asia and Latin America. In all of these situations, reduction in distance to the water source means time saved for women to spend on income-generating tasks, domestic work, and child rearing, or, equally important, on leisure.

There are other benefits, too. More accessible water makes it possible to increase output from home gardens and small animal farming, activities that are commonly the responsibility of women. And the prognosis for child health improves when the amount of water available for washing and bathing increases, the bacteriological quality of drinking water improves, and excreta disposal facilities are built and used (*see* Box 1). So women have a considerable vested interest in seeing water supply improvements successfully implemented and sustained.

It is noteworthy too that, because fetching and carrying water is usually a woman's task (except where money is earned), men typically do not value improved water supplies as highly as women do. In Zimbabwe, for instance, women are willing to pay 40 percent more than men for an improved water supply.

Even if increased participation of women in water supply projects were not an end in itself, therefore, it is a fruitful way of increasing the likelihood of success in such projects. The fact remains that in many societies formidable barriers prevent women from full participation in social, economic, and political life. Carrying water is considered women's work, but when it comes to making decisions and acquiring technical skills, it is usually men who take command.

The Malawi rural piped water project -- overall a very successful project -- exemplifies how women are often treated unless special efforts are made to take account of their importance in this sector. Women provide over half of the large amount of self-help construction labor and over two-thirds of the committee members responsible for tap maintenance. Yet they make up only 10 percent of the important planning committees and repair teams. Lesotho provides a typical example of the consequences of such neglect. Because drawing water is a woman's task, men have few incentives to carry out repairs and many water systems no longer function. It is now recognized that strengthening the institutions responsible for maintaining water systems requires greater inclusion of those community members -- women -- who have a primary interest in sustaining the community's investment.

First, opportunities have to be provided for women to improve their technical skills to perform these tasks. In most countries, only recently have systematic efforts been made to include women in project development, maintenance, and upkeep. The results have generally been encouraging. In Zimbabwe, women manage, use, and maintain communal water points; in an urban slum in Zambia, the women's branch of the political party organized women to improve drainage around public taps; women have been trained as handpump caretakers in Bangladesh, India, Kenya, Lesotho, and Sudan; in Mozambique, women pump mechanics perform alongside, and as effectively as, their male counterparts; in Sri Lanka, women's cooperatives have been set up to assemble and maintain a locally manufactured handpump; in Paraguay, women are trained to manage water systems; women's cooperatives manage communal standpipes and collect money to pay for metered supplies in Honduras, Kenya, and the Philippines. Experience also shows that where women are trained to manage and maintain community water systems, they often perform better than men, because they are less likely to migrate, are more accustomed to voluntary work, and because they are more trusted to administer funds honestly.

The desire to maximize health benefits from improved water supplies also implies a key role for women. The effect of rural water supply projects is substantially greater when hygiene education and sanitation activities are part of the water project. Because of women's central role in household hygiene, food preparation, and childcare, such programs are of necessity *for* women, and, because woman-to-woman communication in such matters is more effective than man-to-woman communication, the programs should usually be conducted *by* women.

Developing Better Ways of Assessing Options, Benefits, and Costs

Improved water supply facilities can bring rural people large benefits in terms of more time, improved health, and better opportunities for production and leisure activities. The benefits are especially important for women, who work long hours and are often extremely poor. This is particularly so in female-headed households, which constitute a large and growing proportion of rural households. This paper has argued that rural people in general and rural women in particular perceive most of these benefits, that they judge the benefits to be large, that they are able and willing to participate actively in improving the quality of their water supplies, and that policies to alleviate poverty must provide support for people to make improvements that they can sustain and that can be widely replicated.

If program planners are to prepare investment projects following the principles outlined here, they need tools to enable them to assess alternatives. For project identification, they need rough screening tools, which can provide insight into the level of water supply service that is likely to be appropriate in a particular type of community and which give a preliminary indication of benefit-cost ratios. For project preparation, more precise tools are needed to determine the willingness to pay for different service levels and to assess the consequences of this information on technology choices (type and number of water points, scale and timing of developments) and financial decisions (user charges, subsidies, and cost recovery targets). Some preliminary efforts have been made to develop such tools, and some further work is needed.

Assessing Benefits and Costs

All decisions are based on some form of benefit-cost assessment, even when the comparison is more intuitive than calculated. The question is not whether to compare benefits with costs, but how.

The benefit most frequently used to justify rural water supply and sanitation investments is health. Indeed, the proclamation of the *International Drinking Water Supply and Sanitation Decade* and the formulation of current rural water supply and sanitation strategies are predicated on the assumption that use of improved facilities will lead to substantial health improvements. In the light of the difficulties of identifying and measuring health effects (*see* Box 1), there are two possible ways of proceeding.

The first procedure, which has been dominant in the sector, is to conclude that benefit-cost analyses are impossible, to assert that projects are justified on health grounds, and to provide a predetermined level of service to the population. The big disadvantage of this approach is that it removes a major discipline from the planning exercise, and leads to the uncritical acceptance of water projects as necessarily "good" and to *ad hoc* procedures for deciding such vital issues as appropriate levels of service.

The second procedure acknowledges the existence of health benefits but uses the other benefits from rural water projects to carry out a partial benefit-cost analysis. In particular, it recognizes that time saving is often a major benefit. Since time saving is also readily perceived and valued by the community, analyses based on such savings provide a useful insight into the likely response of users to suggested alternative improvements. So, it is argued, partial benefit-cost analysis can pro-

Figure 2. Appropriate Levels of Service When Time Savings Are Considered As the only Benefit

		Income	
		Low	**High**
Water availability	**Easy**[a]	No project justified	Yard taps
	Difficult[b]	Standpipes where electricity is available and drilling is expensive Handpumps in all other situations	

a. Traditional sources close; pumping lifts low.
b. Traditional sources far; pumping lifts deep.

vide planners with a helpful screening tool, suggesting in any particular setting what the appropriate level of service may be and giving a lower limit to the benefit-cost ratio.

A method to assess some of the benefits and costs of rural water supply alternatives has been developed by the World Bank Handpumps Project. The method has been used to determine appropriate levels of service under conditions prevailing in different parts of the developing world. The findings, considered in conjunction with other factors common in rural areas (such as the difficulty of assuring a supply of diesel fuel) suggest that the technologies of choice will often be as shown in Figure 2. The results suggest that, where existing sources are close and incomes are low (as in much of Bangladesh, for instance), rural water projects are unlikely to be justified on the basis of time savings alone. (Coincidentally, it is in precisely these settings that the direct health effect of improved water supplies is likely to be small.)

The results also show, however, that where the distance to existing sources is moderate or high, or where incomes are moderate to high, rural water projects are likely to be justified on the basis of time savings alone, without recourse to the difficult-to-measure health benefits. The two other major categories of benefits associated with rural water projects (the perceived benefit of a modern service and health benefits) will generally tend to be higher for standpipes than handpumps, and still higher for yardtaps.

Realizing that there are significant differences between countries in a particular region and between villages in a particular country, it is nevertheless informative to assess what the service improvements appropriate in some different settings might be. Figure 2 suggests that the following may often be appropriate:

- *South Asia and Sub-Saharan Africa*
  - with water readily available: no improvements
  - with electricity available: standpipes
  - without electricity available: handpumps
- *Much of Latin America and Western Asia:* yard taps

It is encouraging that these conclusions fit well with actual situations where careful analysis of the available options has been carried out. In Kenya, for instance, detailed analyses have shown that individual connections are to be encouraged in

Box 12. Models for Assessing Some of the Benefits and Costs of Rural Water Projects

A procedure that allows planners of rural water supply projects to compare time saving benefits with costs is useful in two ways. First, it gives a lower limit to the benefit-cost ratio; second, because time savings are usually among the most important benefits perceived by users, the procedure suggests which level of service villagers are likely to opt for in any particular setting.

A model has been developed by the UNDP-funded World Bank Handpumps Project to perform such a comparison for planners in the early stages of project preparation. The costs of water provided by different water supply systems are compared at the point of use (the home) by adding the costs of piping (for yardtap supplies), the value of time spent hauling water (for handpumps), or a combination of the two (for standpipes). Benefits computed in the model comprise reductions in the time taken to collect water and the value of increased consumption. Net benefits can thus be compared for different service levels in any prescribed setting.

The model serves a number of useful purposes. The Inter-American Development Bank, which requires all borrowers to use a similar procedure in preparing water projects, has found that the partial benefit-cost ratios emerging from such analyses are often substantial and large enough in themselves to justify project implementation, without the need to estimate uncertain health benefits. The model can also be used to assess which level of service is likely to be appropriate in any particular setting. And, finally, sensitivity analyses can be conducted by varying parameters, such as well costs, energy costs, the value of time,[a] and the distance to existing sources. In this way, planners can identify those elements which are likely to be critical in final design decisions and which should therefore be the focus of further data collection and research.

Useful as the model is, it is not intended to make decisions, but only to provide decisionmakers with a rough estimate of the minimum economic return from a project, a rough guide to the most appropriate level of service, and an indication of the most important factors in project design.

a. A particularly problematic element in calibrating the model involves an estimate of the value of the time — usually women's time — saved when water is brought closer to the point of use. Efforts to assess the value of time have been fraught with theoretical and empirical difficulties in both developed and developing countries. With little justification for the differences, planners in development banks have valued time savings at between 33 and 100 percent of wage rates in transport projects, and from 25 percent to 100 percent in rural water projects.

Box 13. Assessing Willingness to Pay for Water and Sanitation Services

There are two ways of determining what water supply and sanitation services people want and are willing to pay for. In the first, "indirect" method, data are assembled on who has actually chosen to install yardtaps, who has opted for standpipe supplies, and who continues to use traditional sources. Inferences are then drawn about the sociological, cultural, and economic factors that may have influenced the choices made. The second, "direct" approach is to ask potential consumers how much they are willing to pay for different types and levels of service and then to draw conclusions about how many households would choose different types of services at different prices.

The indirect method has one big advantage in that data are drawn from actual practices. It also has several disadvantages: the service to be offered (say, a yardtap) may often not exist in the community, there are often large discrepancies between the apparent value of a service and the amount that individuals are actually prepared to pay, and this is a top-down approach, which designs for rather than with the community.

The major, obvious, difficulty with willingness to pay questions is that the answers may be misleading, either because a hypothetical question does not elicit sufficiently serious consideration or because there are perceived advantages to giving false answers. In 1976, the World Bank concluded that use of such surveys in rural water supply was "virtually useless" -- confirming a view that was widely held at the time. More recent experiences in industrialized countries, where the method has been widely used to assess the benefits of public goods, have led to major theoretical and methodological advances in understanding the biases in such direct surveys. It now appears that the method has substantial potential for assessing the demand for public services in developing countries. It is likely that an appropriate method for assessing willingness to pay as part of project preparation activities will include both indirect and direct measures of likely responses to improved services.

"high potential areas" (where water is relatively easily available and incomes are relatively high), whereas lower levels of service (standpipes or handpumps) will be promoted in the less densely populated and poorer "low potential areas."

Useful as the model that is illustrated in Box 12 is, it is important to emphasize its limitations. Some of the information required as input (in particular, the value of time) is impossible to specify with precision in any particular setting. Furthermore, there are several important benefits from rural water supply other than those evaluated by the model. For example, detailed evaluations of U.S. Agency for International Development (USAID) rural water projects in Kenya, Panama, Peru, Tanzania, and Thailand showed that, in addition to time savings, users considered other benefits to be important, including economic benefits (home gardens, domestic livestock, brickmaking, and beer production) and use of types of service (such as piped water) which are considered more attractive and "modern."

The role of the model then is not to make decisions, but only to provide decisionmakers with a rough lower-bound estimate of the economic return from a project, a sense of the level of service likely to be appropriate in a community, and an indication of factors (such as the income and drilling costs) that are important in project design.

Assessing Willingness to Pay

Just as planners' decisions are implictly based on some form of benefit-cost assessment, so too are the decisions of villagers. Thus, there is a link between the benefits included in the partial benefit-cost model and the willingness of people to pay for an improved water supply. In any specific setting, however, it will be impossible to calibrate the partial benefit-cost model precisely (mainly because of the difficulty of assigning a monetary value to time savings) or to include all the benefits perceived by villagers.

There are other important limitations to any procedure that does not require direct consultation with the community involved. Colombia, Malawi, and other countries have shown that active community involvement in planning, design, construction, and maintenance is a key element of successful rural water supply projects. It follows that a methodology for assessing villagers' willingness to pay for different levels of service should involve active consultation, not simply analytical inferences. Box 13 discusses the potential of the alternative approaches in more detail.

3. How the International Community Can Help

Until comparatively recently, the aid community has been sharply divided on rural water strategy. To simplify a complex debate: some institutions have argued that the financial viability of the sector must be a primary concern, while others have maintained that the highest priority should be the rapid expansion of services to meet the basic needs of large populations.

It is now clear to donors and developing countries alike that, far from conflicting, the two perspectives are complementary, and that sustainability and replicability are closely related facets of an appropriate strategy for developing the sector. The past few years in particular have seen unparallelled collaboration among multilateral and bilateral assistance agencies active in the water supply and sanitation sector. Many of the policies advocated here have already been included in joint statements issued by major donors following consultations in Europe, Asia, Africa, and the Americas.

These recent developments raise hopes that we may be able to move away from the situation in which within a single country there may be as many rural water supply and sanitation policies as there are donors. In the past, so-called "donor coordination" has often failed, but the climate has now improved. Widespread dissatisfaction with early progress in the Water Decade prompted the donor consultations, and there is now considerable convergence of opinion on how to improve assistance in the sector (Box 14).

Developing countries, too, are endorsing the new approaches. A recent seminar in Abidjan, Côte d'Ivoire, resulted in an agreed strategy, endorsed by 30 Sub-Saharan African countries and 15 external assistance agencies, which emphasized the need for progress based on:

- Involvement of community members in the planning, design, financing, and maintenance of improved water supplies, with women's groups taking a leading role.
- Use of public and private sector resources to provide initial training and necessary long-term support, so as to create the environment in which community management can function successfully.
- Technology choices based on affordability and sustainability.

Progress in any particular country will now depend on these principles being reflected in national policy and on the development, field-testing, and scaling-up of methods to put the principles into practice (with appropriate resource commitments). A vital task in each country is a policy dialogue between external financing agencies and national governments. Once agreement has been reached on the broad outlines of a country strategy, governments and agencies together can prepare sound sector-wide and project-specific investment projects.

Particularly in the early stages of implementation, this will require substantial investments of staff time by donor and borrower alike. Institutional analyses will have to be performed to specify the roles of different public sector, private sector, and community institutions. Technical, human resource, and financial constraints will have to be identified. And realistic targets will have to be set.

Project preparation too will take more time than it generally has in the past. Involvement of the community in deciding what services will be provided and how they will be funded and maintained is a central element of the new approach. This process, for which validated practical methods still need to be developed, is necessarily more time consuming than the conventional top-down methods widely used at present in the sector.

In broad terms (for which there are both positive and negative exceptions), it is likely that progress will continue to be relatively fast in much of East Asia, moderate in much of South Asia and Latin America and slow in much of Sub-Saharan Africa.

In conclusion, through a variety of efforts by external agencies, developing country governments, and communities themselves, there has been significant experience with the key elements of the approach outlined in this report. While it will not be easy to make the necessary changes, accumulated experience with this approach suggests that a concerted effort along these lines would substantially improve the lives of many women, children, and men in the rural areas of developing countries. The emerging consensus on these issues among external funding agencies and national governments suggests that such an effort might indeed be possible.

Box 14. Prospects for Donor Coordination

Examination of the systemic effects of a multiplicity of donors working in a particular country points consistently to a need for better coordination. Recent reviews of the performance of aid projects deal at length with the need for, and prospects of, better coordination among donors. Among the issues raised are:

- Aid can only be as effective as the policy, economic, and administrative environment in which it operates. Aid therefore has to be concerned with creating the fundamental conditions for its own effectiveness.
- If aid is to make a broad and sustained contribution, it must be concerned not only with the proper selection, design and implementation of individual projects, but also with the support of broader sectoral and national efforts and policies.
- With many donor agencies operating in a given sector or subsector, project and equipment types proliferate (for example, 18 makes of pump are supplied through aid for Kenya's rural water supply, each with different requirements for installation, maintenance, parts, and training).
- A multiplicity of donors, the complexity of their separate procedures, and the requirements of project implementation and monitoring can put intolerable burdens on the recipient's administration.
- The recipient's budgetary process may be inadequate to cope with the demands of multiple uncoordinated aid activities for domestic resources to complement the external ones supplied by donors.

These problems abound in countries of slender administrative means, particularly in Sub-Saharan Africa.

Until recently, donors and recipients resisted coordination for many reasons. On the recipient side, governments have often preferred to retain an ability to play one donor off against another, and have been apprehensive about donors ganging up on them if aid coordination were to be effective. On the donor side, bilateral agencies were keen to pursue national commercial and political objectives that might not be served by coordination with other donors. Donors also were concerned that they might not be able to agree with one another on significant policy issues.

In the water supply and sanitation sector, both donors and recipients are now more ready to appreciate that the gains from past lack of coordination are small in comparison with the costs. Improved coordination in any country must start with, and be controlled by, the recipient, helped if necessary through technical assistance.

For the decade ahead, there is reason both for concern and for hope: concern because many developing countries have yet to lay a foundation that will permit rural people to meet their needs for improved water supplies; hope because some countries have shown how this can be done. There is an additional reason for hope, too, because there is an important resource that has not yet been brought to bear fully, a resource of tremendous potential, energy, and ingenuity, if only it can be effectively harnessed: the *people* themselves in the unserved communities. No one can predict how much can be achieved if local people's aspirations, initiatives, persistence, and self-interest are given freer rein to bring about faster progress; but experience suggests that the possibilities are vast.

Bibliography

Confidential World Bank documents and the following references were used in developing this report.

Arlosoroff, S., G. Tschannerl, D. Grey, W. Journey, A. Karp, O. Langenegger, and R. Roche. 1987. *Community Water Supply: The Handpump Option.* Washington, D.C.: The World Bank.

Birdsall, N., and F. Orivel. 1983. "Willingness to Pay for Health and Water in Mali: Do WTP Questions Work?" Population, Health and Nutrition Department, World Bank, Washington, D.C.

Briscoe, J., M. Chakroborty, and S. Ahmed. 1981. "How Bengali Villagers Choose Sources of Domestic Water," *Water Supply and Management*, 5, 165-81.

Briscoe, J., R. G. Feachem and M. M. Rahaman. 1986. *Evaluating Health Impacts: Water Supply, Sanitation and Hygiene Education.* Ottawa: IDRC Press.

Cairncross, S., and J. Cliff. Forthcoming. "Water Use and Health in Mueda, Mozambique." *Transactions of the Royal Society of Tropical Medicine and Hygiene.*

Cairncross, S., and R. G. Feachem. 1983. *Environmental Health Engineering in the Tropics.* London: Wiley.

Calegary, S., J. Schwartz, and R. Verspyck. 1987. "New Lending Opportunities and Lending Tools: Water Sector Adjustment Loans," Water and Urban Development Department, World Bank, Washington, D.C.

Churchill, A. A. 1987. *Rural Water Supply and Sanitation: Time for a Change.* World Bank Discussion Paper No. 18. Washington, D.C.

Coopers and Lybrand Associates Limited. 1979. "Egypt Provincial Water Supply Project: A Report on a Survey of Consumers' Attitudes to Potable Water Supplies." London.

Dworkin, D. 1980a. "Kenya Rural Water Supply: Programs, Progress, Prospects." USAID, Washington, D.C.

_____. 1980b. "The Potable Water Project in Rural Thailand." USAID, Washington, D.C.

_____. 1980c. "Rural Water Projects in Tanzania: Technical, Social, and Administrative Issues." USAID, Washington, D.C.

_____. 1982. "Community Water Supply in Developing Countries: Lessons from Experience." USAID, Washington, D.C.

Elmendorf, M., and P. Buckles. 1978. "Socio-cultural Aspects of Water Supply and Excreta Disposal." Public Utilities Report No.13, Infrastructure and Urban Development Department, World Bank, Washington, D.C.

Esrey, S. A., R. G. Feachem, and J. M. Hughes. 1985. "Interventions for the Control of Diarrheal Diseases among Young Children: Improving Water Supplies and Excreta Disposal Facilities." *Bulletin of the World Health Organization* 63(4):757-72.

Feachem, R., E. Burns, A. Cairncross, A. Cronin, P. Cross, P. Curtis, M. Khan, D. Lamb and H. Southhall. 1978. *Water, Health and Development: An Inter-Disciplinary Evaluation.* London: Tri-Med Books.

Freedman, J., and J. Listorti. 1985. "World Bank Rural Water Supply and Sanitation Lending, 1974-1984: a Retrospective Review." Water Supply and Urban Development Department, World Bank, Washington, D.C.

Government of India, Ministry of Works and Housing. 1981. "National Master Plan -- International Drinking Water Supply and Sanitation Decade." New Delhi.

Government of India, Ministry of Works and Housing. 1983. "Report of the Working Group on Operation and Maintenance of Drinking Water Supply Systems." New Delhi.

Government of India, Planning Commission. 1980. *Accessibility of the Poor to Rural Water Supply.* New Delhi.

Kalbermatten, J. M., D. S. Julius, C. G. Gunnerson and D. D. Mara. 1982. *Appropriate Sanitation Alternatives: A Planning and Design Manual.* Baltimore, MD: Johns Hopkins University Press.

Mora Ramirez, J., and O. Lopez Orosco. 1976. "Participacion comunitaria y saneamiento basico rural en Colombia." *Carnets de l'Enfance* (34): 89-101.

Operations Research Group. 1984. "Functioning of Handpumps and Water Uses in Selected Rural Areas: Summary Report" UNICEF, Bhubaneswar, India.

Raditski, M. 1986. "Swedish Aid to Kenya and Tanzania: Its Impact on Rural Development." Development Research Division, World Bank, Washington, D.C.

Roche, R., and W. Frederick. 1987. "Cost-benefit Model for Community Water Supply: Technology Choice." Infrastructure and Urban Development Department, World Bank, Washington, D.C.

Roth, G. 1987. *The Private Provision of Public Services in Developing Countries.* Washington, D.C.: World Bank.

Roy, S. 1984. "A One-Tier System: The Tilonia Approach to Handpump Maintenance." *Waterlines.*

Saunders, R. J., and J. J. Warford. 1976. *Village Water Supply: Economics and Policy in the Developing World.* Baltimore, MD: Johns Hopkins University Press.

Selowsky, M. 1979. *Who Benefits from Government Expenditure? A Case Study of Colombia.* New York: Oxford University Press.

Swedish International Development Agency. 1982. *Socio-economic Survey of Water Use in Kenya.* Nairobi.

USAID. 1981. *Korean Potable Water System Project: Lessons from Experience.* Washington, D.C.

Warner, D., J. Briscoe, C. F. Hafner, and B. Zellmer. 1986. "Malawi Self-help Rural Water Supply Program: Final Evaluation." WASH Project, Washington, D.C.

Wellin, E. 1983. "Village Water Systems in Peru: Community Development or Dilemma?" Paper presented at the XIth International Congress of Anthropological and Ethnological Sciences, Vancouver.

White, G. F., D. J. Bradley, and A. U. White. 1972. *Drawers of Water: Domestic Water Use in East Africa*, Chicago: University of Chicago Press.

Whittington, D. J. Briscoe, and X. Mu. 1987. "Willingness to Pay for Water in Rural Areas: Methodological Approaches and an Application in Haiti." WASH Project, Field Report No. 213, Washington, D.C.

Wijk-Sijbesma, C. 1979. *Participation of Education in Community Water Supply and Sanitation Performances: A Literature Review*, The Hague: International Reference Centre.

____. 1985. "Participation of Women in Water Supply and Sanitation: Roles and Realities." The Hague: International Reference Centre.

World Bank. 1976. "Measurement of the Health Benefits of Investments in Water Supply." Report PUN 20. Infrastructure and Urban Development Department, Washington D.C.

____. 1984. *World Development Report 1984.* Washington, D.C.

____. 1985. "Rural Water Supply and Sanitation: Possibilities for Collaboration with Non-governmental Organizations. Part One: Africa." Water Supply and Urban Development Department Technical Note. Washington, D.C.

World Health Organization. 1984. *Guidelines for Drinking Water Quality. Volume 1: Recommendations.* Geneva.

World Water. 1986. "Water Decade Review."

Yucel, N. C. 1975. *A Survey of the Theories and Empirical Investigations of the Value of Travel Time Savings.* World Bank, Staff Working Paper No. 199. Washington, D.C.

Zimbabwe, Republic of. 1985. *National Master Plan for Rural Water Supply and Sanitation.* Harare.